SCIENCE AND THE SWASTIKA

SCIENCE AND THE SWASTIKA

adrian weale

First published 2001 by Channel 4 Books
an imprint of Macmillan Publishers Ltd
25 Eccleston Place London SW1W 9NF
Basingstoke and Oxford

www.macmillan.com

Associated companies throughout the world

ISBN 0 7522 1931 6

Copyright © 2001 Adrian Weale

The right of Adrian Weale to be identified as the author of this work has been asserted by him in accordance with the Copyright, Designs and Patents Act 1988.

All rights reserved. No part of this publication may be reproduced, stored in or introduced into a retrieval system, or transmitted, in any form, or by any means (electronic, mechanical, photocopying, recording or otherwise) without the prior written permission of the publisher. Any person who does any unauthorized act in relation to this publication may be liable to criminal prosecution and civil claims for damages.

9 8 7 6 5 4 3 2 1

A CIP catalogue record for this book is available from the British Library.

Design and typesetting by Jane Coney
Printed by Mackays of Chatham plc, Chatham, Kent

While every effort has been made to trace copyright holders for photographs featured in this book, the publishers will be glad to make proper acknowledgements in future editions in the event that any regrettable omissions have occurred at the time of going to press.

Page 1: (top) AKG London, (bottom) David Irving/Focal Point Publications
Page 2: (top) David Irving/Focal Point Publications, (bottom left) Max Planck Gesellschaft, (bottom right) Niels Bohr Archive, Copenhagen
Page 3: (top) Max Planck Gesellschaft, (bottom left) Ullstein Bilderdienst, (bottom right) Bundesarchiv Document Centre, Bild 183/1991/0904/502 (Alfred Ploetz, 1931)
Page 4: Bildarchiv Preussischer Kulturbesitz
Page 5: (top) reproduced courtesy of United States Holocaust Museum, (centre left) Bildarchiv Preussischer Kulturbesitz, (bottom right) Besançon
Page 6: Auschwitz Museum
Page 7: Ullstein Bilderdienst
Page 8: (top) Bildarchiv Preussischer Kulturbesitz, (bottom) Yad Vashem

FILM & TELEVISION PRODUCTION

This book accompanies the television series 'Science and the Swastika' made by Darlow Smithson for Channel 4.
Executive Producer: David Darlow
Series Producer: Dunja Noack
Producer of Programmes 1 & 3: Saskia Baron
Producer of Programme 2: Paul Sen
Producer of Programme 4: Fisher Dilke

This book is sold subject to the condition that it shall not, by way of trade or otherwise, be lent, re-sold, hired out, or otherwise circulated without the publisher's prior consent in any form of binding or cover other than that in which it is published and without a similar condition including this condition being imposed on the subsequent purchaser.

IN MEMORY OF
MY FATHER:

DR K.E. WEALE FIC

ACKNOWLEDGEMENTS

This book would not have been possible without the hard work of a number of people, not the least of whom are the team at Darlow Smithson responsible for researching and producing the 'Science and the Swastika' TV series. I would like to mention, in no particular order: David Darlow, Marion Lacey, Saskia Baron, Shaun Whiteside, Paul Sen, Jackie Murphy, Fisher Dilke, Andrea Laux and Dunja Noack.

Much of the background material has been culled from visits to the Public Record Office in Kew, the Haldane Library at Imperial College, the Imperial War Museum Library, Kensington Library, the Berlin Document Center and correspondence with the United States National Archives and the Bundesarchiv in Germany.

At Channel 4 Books I am happy to thank Charlie Carman and Verity Willcocks for their sterling efforts; and also Andrew Lownie for his fearless agenting.

I must also thank Richard Dawes for his concise and sympathetic copyediting, which has taken much of the pain out of preparing the manuscript for print.

At home, I must also thank Anne Grosset for looking after Robert and Ivo, and, of course, my wife Mary, for not getting too irritable as I sat up until one in the morning and got up at five to finish the book.

CONTENTS

introduction

In the 1930s Germany was arguably the most scientifically and technologically advanced country on earth. For more than fifty years German scientists had been at the forefront of progress in fields such as theoretical nuclear physics, psychiatry, engineering, avionics, chemistry and sociology; in 1936 the world's first television broadcasts were made from the Berlin Olympic Games; in 1939 the first epidemiological study linking smoking with lung cancer was published in Germany. In fact, by the time the Second World War broke out, German scientists were well ahead in the unofficial league tables of Nobel laureates and, as an example, German physicists had wholly or partly won eleven of the thirty-seven physics prizes which had been awarded so far.

But, in a matter of a few years, all this was to change: by 1945 Germany was physically in ruins and German science had suffered a series of devastating blows. Its theoretical physicists had failed to get close to building an atomic bomb or even a working nuclear reactor; the engineers, despite making some striking technical advances, had been unable to produce weapons of sufficient quality in sufficient quantity to prevent Germany's defeat; while German doctors and biologists stood accused of participating in, and to some extent instigating, a system in which more than 400,000 Germans were compulsorily sterilized, 70,000 so-called 'incurables' had been murdered in gas chambers (by doctors and nurses) and what advances had been made were often the result of obscene experiments on unwilling and sometimes unwitting subjects. The reasons why Germany had been so successful scientifically are relatively easy to explain; the reasons for the catastrophic decline are somewhat more complex.

The great engine of scientific advance in the eighteenth and nineteenth centuries was the industrial revolution, when the countries of Europe and later America began to switch from an essentially rural and agricultural economy to one based on the manufacture of mass-produced goods. This originally started in Britain but was soon followed in the other nations of northern Europe, including Prussia and the other states and statelets inhabited by the German peoples. In Britain and elsewhere in Europe most technical improvements and modernization were carried out within, and at the behest of, commercial concerns and for commercial reasons. Industrial modernization required technological advances – important examples are steam engines and electric light – but it also needed a hinterland and underpinning of scientific understanding.

Until the eighteenth century science was the province of 'natural philosophers': men who applied their observations of everyday life and phenomena – like the fact that apples fall from trees – in an attempt to understand the fundamental nature of the world around them. A few of the natural philosophers were men of towering genius like Isaac Newton or Galileo, but the majority were less successful, and science, as it was practised by metallurgists, pharmacists, alchemists and other even less plausible charlatans, was based on a ramshackle collection of bizarre doctrines: seven metals were each associated with the seven known planets; base metals could be transmuted into gold; the four elements were declared to be water, air, fire and earth; and, as an example of applied science, underground mines were occasionally sealed up and left to allow time for further growth of the minerals they contained.

What changed this situation was the development of scientific method. This is the process by which a hypothesis is formed on the basis of known data and then tested and proved (or refuted) by experiment and observation. In theory, at least, this should bring about a process of evolutionary progress in scientific knowledge as newly discovered facts point the way towards further study, and in many respects this is what happened. Scientific methodology provided an academic basis for further study in fields which had hitherto proved impenetrable.

It so happened that science as an academic pursuit was taken up with more vigour at German universities than elsewhere. In England, Oxford and Cambridge, the oldest universities, resisted the arrival of scientists, according to Professor Max Perutz, the Nobel Prize-winning biochemist: '[German Universities have a] long tradition. I think science started at German universities much sooner than [in England], you know. Here the classical establishment tried to keep them out and so there was a great deal of excellent science in Germany in the nineteenth century when there was far less here.'

Germany's industrial revolution followed the one in Britain and, as a consequence, German industrialists were able to avoid many of the pitfalls that their British predecessors had experienced. Other spurs which motivated German industry and business included the lack of a substantial overseas empire, a situation which forced them to compete more fiercely to establish markets for their goods, and the process of German unification, which took place in the middle of the nineteenth century.

Unification, steered by the nationalist Chancellor Otto von Bismarck, also introduced a 'statist' concept of government that was very much in keeping with the Prussian intellectual tradition. In essence this held that the individual had a duty to the state, while downplaying the notion of individual rights, and justified the very hierarchical nature of German society. Insofar as pre-First

World War German society was at all democratic, it was a consultative democracy built around a highly authoritarian core.

As a people the Germans have always shown enormous respect for practical achievements and academic success, and this is combined with a consciousness of status bordering on the fanatical. A businessman or academic visiting Germany will almost always be addressed, under formal circumstances, as 'Herr Direktor', 'Herr Ingenieur' or even 'Herr Professor Doktor'. The concern for establishing status brings with it a devotion to hierarchy and respect for authority which is also without parallel.

Unified Germany was, in addition, strongly militarist. The unification process had required the fighting of a series of short campaigns and these were followed by the Franco-Prussian War of 1870–1, during which the highly efficient, modern Prussian-based German army had resoundingly defeated a larger but considerably less modern French force. Towards the end of nineteenth century Kaiser Wilhelm, head of state of the unified Germany, decided to attempt to challenge British maritime supremacy by entering into a naval warship-building race in the hope that he would force concessions towards his colonial ambitions. Together the demands of the German army and navy created both a major armaments industry, based around giant industrial combines like Krupp and Mauser, and a military-scientific infrastructure.

Perhaps the most important single factor in pushing Germany to the forefront of science in the first half of the twentieth century came with the foundation of an elite scientific research organization. The leading force in this was Adolf von Harnack, a theological scholar by training, who was one of Germany's most influential figures in the promotion of natural sciences.

In 1910 Harnack collaborated with Emil Fischer, Walther Nernst and August von Wassermann to establish a new organization dedicated to the service of German society through the advancement of scientific research. Their original proposal for this organization emphasized basic principles for the cooperation of science, industry and government and included the creation of a series of elite research institutes. The proposal underscored the importance of complete research freedom for its members, structurally supported by the award of lifetime contracts to a few carefully chosen research directors.

Before long Harnack and his colleagues had gained the personal backing of Kaiser Wilhelm II, who also provided a name for the Kaiser Wilhelm Society (Kaiser Wilhelm Gesellschaft or KWG). They then successfully solicited the support of many of the nation's leading industrialists and bankers. The KWG was formed with the then astronomical budget of 10 million marks.

Harnack was elected President and the first of the KWG's institutes opened in 1911 in Berlin. Known for his liberal views, Harnack had his own basic hiring principle: scientific ability first. He recruited some of the country's most important scientific figures for the KWG's first institutes, including Richard Willstätter, Otto Hahn, Lise Meitner, Ernst Beckmann, Fritz Haber, August von Wassermann and Albert Einstein.

Very soon after this, however, Europe descended into crisis. In July 1914 Austria-Hungary attacked Serbia in response to the murder of a member of the Austrian ruling family. Russia gave such strong indications of support for its Serbian ally that Germany, which was allied to Austria-Hungary, declared war on Russia and on France, which was allied to Russia. Germany had a well-prepared plan for war with France, designed to gain a quick decision in the West so that the inevitably more protracted war with Russia could be resolved over a longer time. The German plan involved an attack through neutral Belgium, thereby ensuring that Britain, which had long acted as guarantor of Belgian neutrality, would become involved in the war. By mid-August 1914 many of Europe's nations were at one another's throats, with Germany and Austria-Hungary, the 'Central Powers', ranged against the British Empire, France and Russia, the 'Triple Entente'.

A series of battles in north-east France led to tactical stalemate and the war on the Western Front soon became a bloodbath, as both sides settled into a well-constructed trench system which stretched from the Swiss frontier in the south to the North Sea coast of Belgium.

Fighting in the first year of the war was catastrophically expensive, in men and munitions, for both sides, and in an attempt to break the stalemate the Germans made what was to be the major military-scientific advance of the war. In the afternoon of 22 April 1915 a slight northerly breeze sprang up, blowing gently from behind the German lines into the faces of Allied soldiers in positions around the village of Langemarck near Ypres. At five o'clock three red rockets were fired into the air, signalling the beginning of a large artillery barrage, but, as the French troops on the receiving end of the gunfire took shelter, many of them noticed an odd greenish-yellow cloud apparently rising from the German trenches. German engineers were opening the valves of 6,000 cylinders of chlorine gas, releasing 165 tons of it in a dense cloud that rolled towards the French lines. Heavier than the surrounding air, the chlorine formed a blanket between 5 and 7 feet high, and in no more than a minute or two had crossed no-man's-land and broken into the French trenches.

For a few moments the bewildered French soldiers wondered what was happening to them as the dense cloud blocked their view of other men standing with them, then the gas began to affect them, attacking the mucous membranes in their throats, bronchial tubes and lungs. The inflammation this started caused their bodies to produce massive amounts of a yellow fluid, filling their windpipes and lungs. But their attempts to breathe caused them to suck more chlorine into their lungs and very soon they were drowning in their own mucus. Some of the last to die probably saw German infantrymen advancing carefully through the thick cloud, each wearing a crude respirator of cotton gauze tied across his face.

This first use of poisonous gas – in which at least 5,000 French soldiers were killed – was followed two days later by an attack in the British sector which accounted for a further 5,000 dead and twice as many wounded, and was to lead to a chemical arms race between Germany and the Allies which would last throughout the war. In fact gas became a principal weapon for Germany. Deprived of raw materials by the Allied blockade, it was forced to fall back increasingly on its chemical industry. Fritz Haber, the chemist who had devised the idea of using chlorine as an offensive weapon, was also responsible for keeping Germany in the war by his discovery of a method for producing ammonia by fixing nitrates from the air, which allowed the synthesis of explosives.

The war continued with a series of attempts by both sides to break the stalemate but no decisive result was achieved until the autumn of 1918, when Germany, exhausted by four years of war on two fronts (Russia had actually sued for peace after the Bolshevik coup d'état in 1917), was finally defeated in the west by the combined British and French armies together with the newly arrived Americans. The cause of Germany's defeat was the brutal attrition that it had suffered, along with the realization by its senior generals that the arrival of the American army, backed by the almost limitless resources of the USA, had decisively tipped the balance in favour of the Allies.

German defeat in the field was followed by the Treaty of Versailles, signed in June 1919, which demanded that Germany accept guilt for starting the First World War, severely restricted the size of the German army (and banned Germany from acquiring an air force) and imposed upon Germany the payment of reparations for damage inflicted on Belgium and France. Additionally, Germany was required to cede territory, including Alsace and Lorraine, to France, and a corridor allowing free passage to the sea to the otherwise landlocked new state of Poland.

The Treaty of Versailles effectively guaranteed years of political turmoil in Germany. The French government, in particular, was in no mood to compromise over reparations payments, which would cripple Germany's economy for many years to come and contribute to the ruin of the German middle classes; while the majority of Germans, who broadly believed that their country had not been militarily defeated but stabbed in the back by treacherous politicians at home, were dismayed by its harshness. One of the many consequences of Versailles was to provoke and stimulate the rise of a plethora of extreme right wing and ex-servicemen's groups and also to make them appealing to a much broader range of society than hitherto. Meanwhile restrictions on German military capability – for example, the army was to be no larger than 100,000 men – also led to secret research programmes and a re-gearing of the German technical development system to work at a more covert level.

Probably the most telling effect of the First World War and the turmoil that followed it was the brutalization of German society. Many Germans came to believe that the best in their society had died at the front, and there was a strong urge for renewal. To a great extent, this explains the rise of the Nazi Party in the years after the war.

National Socialism – Nazism – was the German variant of fascism, the political creed which had grown up in Europe in response to the war and the subsequent growth of communism. One of the key factors in understanding fascism is that, unlike communism, it does not have a specific political programme, but rather defines a group of people with a shared set of beliefs, methods, psychological traits and common enemies.

The most common element of fascism, and therefore Nazism, was fear. In particular this meant fear of communism, or Bolshevik revolution or insurrection, but also of outsiders generally, whether they came from a different country, race or social class or because they looked or behaved markedly differently. Nazism was attractive to the broad German middle classes because it played to their fears and promised to protect them from all such threats. All fascist groups have been ultra-nationalistic, claiming to believe that their nation is endowed with special qualities or a special mission to which the individual must subordinate himself, but all have purported to offer advancement and security against competitors.

The Nazis also liked to think of themselves as radical and modern. Although they are usually categorized as being on the political right, there is a degree of truth in the idea that they can, in certain ways, be more closely identified with the left. They were hostile to conservative or reactionary bodies like the church and the armed forces, which they regarded as backwards-looking. And what could stand in greater contrast to these than science? There is no doubt that Nazism embraced aspects of science which were congenial to it. In the early part of the twentieth century this included eugenics, the science of race improvement, which had grown up towards the end of the previous century and was based on a poorly understood fusion of Darwinian evolution and Mendelian genetics.

Eugenics suited the Nazis well: even when not specifically racist in the sense of promoting hatred or division, it implied a system of racial rank. Not surprisingly, the eugenic theories which most appealed were those which suggested that 'Germanic', 'Nordic' and/or 'Aryan' peoples represented the top of the evolutionary ladder, while the Mediterranean peoples, Slavs and Jews were some way down the scale, the bottom of which was propped up by Africans. For the average Nazi, such notions were self-evident, and the idea that society could be 'scientifically' characterized as a racial struggle fitted in well with the basic psychology of the members of the movement. Indeed in 1934 a senior Nazi official in Bavaria announced that National Socialism was nothing more than 'applied biology'.

But if Nazism embraced science – or at least those aspects of it which it could understand – science also embraced Nazism. Most scientists were afflicted by the same anxieties that troubled the rest of German society, but 'Aryan' scientists were among the first to benefit from Nazi rule when, in 1933, the Jews were expelled from the civil service and universities. Distasteful as this undoubtedly was to many scientists, it also led to widespread and unexpected promotions as the 'Aryans' moved in to fill the vacancies; and in the frenzied atmosphere of rearmament and racial cleansing in Germany, many of them were able to acquire a status they would not otherwise have had.

Thus the 1930s saw German science as the bearer of a long and successful tradition, about to embark on a relationship with a regime that held it in high regard and saw in science a possible way of purging society of undesirable elements and so renewing itself. Science would give Nazism its theoretical and technological underpinning; Nazism would pass on to science its fundamental lack of moral restraint and certainty that the end justifies the means.

It was an equation which was to have disastrous consequences.

CHAPTER ONE

improving the stock

For at least one young drifter the outbreak of war in August 1914 was a moment of profound joy: 'For me these hours came as a deliverance from the distress that had weighed upon me during the days of my youth... I sank down to my knees to thank heaven for the favour of having been permitted to live in such a time.'

The young man, an Austrian living in Munich called Adolf Hitler, immediately volunteered for service with the German army and was delighted to be accepted into the 16th Bavarian Reserve Infantry Regiment. After a couple of months' training, the regiment was sent to the Western Front in time to take part in the first battle of Ypres in October 1914, and Hitler stayed with it until he was wounded two years later: his first return to Germany since departing for the front in 1914.

Until the war came, his life had been a slow but steady descent from middle-class prosperity, as the son of a reasonably well-to-do Austrian Imperial Customs official, to a state of near vagrancy, living off a small pension he received as the orphan of a civil servant and whatever money he could make selling sketches and postcards on the streets of Vienna and later Munich. In the German army, however, Hitler found structure, discipline and purpose. Unlike many of his fellow front soldiers, of every side, he wasn't repelled by the death and destruction he saw around him: to some degree he revelled in it, for the whole experience seemed to him to be toughening his mind as well as his body.

In the spring of 1918 Hitler's dreams of martial glory appeared to be about to reach their apotheosis. After a winter of hunger and discontent during which workers in Germany had attempted to hold a general strike, the new Bolshevik government in Russia had finally agreed to discuss peace and had agreed to

Germany's crushing terms in the treaty of Brest-Litovsk at the beginning of March. A few weeks later, having concentrated its forces in the west for a last roll of the dice, the Imperial German army launched its final great offensive, driving the British, French and American armies back to within forty miles of Paris. Corporal Hitler, awarded the Iron Cross, First Class, on 4 August, was convinced that victory was within Germany's grasp; he was wrong. On 8 August the British counter-attack at Amiens thumped into the exhausted, strung-out German army and an orderly retreat slowly turned into a rout. For the first time in the war, British soldiers on the advance saw that their German counterparts had begun to throw away vital items of equipment in their haste to get away from the fighting unencumbered. Very soon the German High Command was secretly extending peace feelers to the Allies: it could not continue.

For Hitler this was inexplicable: as far as he was concerned the German army wasn't nearly defeated and, after being caught in a British gas attack, temporarily blinded and hospitalized at the beginning of October, he did not see the final collapse, literally or metaphorically. But as he lay in hospital, unable to comprehend what had overtaken Germany, the idea of attempting to reverse the personal and national humiliation of defeat was implanted deep in his psyche. On his release from the hospital after the armistice had been signed, he made his way through a country unrecognizable in the convulsions of defeat. The old order which had sustained Germany through unification and Kulturkampf – the period in the late nineteenth century when Bismarck largely succeeded in subordinating the country's Catholic Church to the state – was being violently swept away. The Kaiser had abdicated; social democrats had taken control of the central government; and communist and socialist revolutionaries were attempting to impose 'workers and soldiers' councils in the cities and towns. In Munich Hitler found the barracks of his regiment in a chaotic state under the authority of a committee of junior soldiers. Unable to stomach this, he volunteered for service as a guard at a POW camp, where he stayed until March 1919.

Hitler returned to Munich just in time for the Soviet coup d'état in Bavaria led by a small group of Russian agitators, two of whom were Jewish, and was a witness to its violent crushing by the army and the Freikorps militia (which included a young man named Werner Heisenberg, whom we shall meet later). Talent-spotted for his intense loyalty and gift of the gab, Corporal Hitler was sent by the army to undertake a short political indoctrination course at Munich University. After this he was deployed to a camp for returning soldiers as part of an 'enlightenment squad', where his role was to ensure the men developed the correct perspective on recent events in their homeland.

Success in this task led to greater responsibility and in the late summer of 1919 he was given the job of liaison officer between the army and the bewildering number of intensely right-wing parties and factions which had sprung up throughout Bavaria in response to the activities of the leftist revolutionaries. It was here that he found his métier. On 12 September he was sent to make a report on a group which had formed around a former locksmith and a sport journalist, calling itself the German Workers' Party. During the discussions that evening the suggestion was made that Bavaria should secede from Germany and seek union with Austria. Unable to contain himself, Hitler jumped in, angrily denouncing the idea and waxing eloquently against the speaker. Impressed, the leaders of the small group invited him to return.

He did so, and two days after his second visit to a German Workers' Party meeting he accepted the committee's invitation to join the party as member responsible for propaganda and recruitment. He threw himself into this role with enthusiasm and within a month had organized a public meeting attended by over a hundred people. Spurred on by this, in February 1920 he managed to persuade nearly 2,000 to pack into the Hofbräuhaus in Munich to hear him speak. Here he faced down the noisy opposition from his rowdy audience, changed the name of the organization to the National Socialist German Workers' Party, or Nazi Party, and presented the party's twenty-five-point plan to solve Germany's ills. It was at this point that he grasped where his destiny lay: as a politician and orator. A few months later he was discharged from the army and set off down the path that was to lead, slightly less than thirteen years later, to the Chancellorship of Germany and the disaster of the Third Reich.

Hitler immediately struck a chord in post-First World War Bavaria, not only with the ex-soldiers and street-corner toughs who comprised the early membership of the Nazi Party, but with a much wider audience as well. Hitler's initial appeal was to the ex-soldiers like himself who firmly believed that they had not been defeated in the war, but stabbed in the back by the socialists, Bolsheviks, Jews, capitalists and speculators who had then tried to take over the country while the heroes of the armed forces were at the front. Incidentally, in doing so he ignored the reality that many of the revolutionaries were themselves soldiers. He had a slightly slower-burning impact among the lower middle classes, who, like him, believed that their supposed class and status gave them a claim on society that wasn't being fulfilled.

From the late nineteenth century an increasing division had become evident in Germany's middle classes. The upper middle class of professionals, successful businessmen and high-level civil servants had increasingly come to be identified, financially and socially, with the aristocracy and 'ruling' class. But the lower middle class of small farmers and businessmen, shopkeepers and, above all, the great army of white-collar clerks, low-level officials, teachers, civil servants and junior managers, had, in the most modern industrialized country in Europe, come under the twin pressures of large-scale corporate capitalism from above, and organized labour from below. Even before the outbreak of the First World War this had produced a move towards a right-wing radicalism with undertones of nationalism and anti-Semitism. Germany's defeat, and the upheavals which followed it as the old order collapsed, contributed even more to the dislocation of the lower middle class: small businesses failed in the turmoil and the value of hard-accumulated savings were wiped out by inflation. The Nazis' message was that this was the work of the Jews and communists – for Hitler the two words were virtually synonymous – not the inevitable result of German expansionist nationalism, and this resonated as much with the struggling lower middle class as it did with the bewildered former *Frontsoldaten*.

But, by a tragic twist of fate, the simple-minded bigotry and scapegoating practised by the Nazis intersected at this point in German history with a quite different and, in some superficial respects, more elevated strand of thought, and the combination was ultimately to produce the unimaginable horrors of the Holocaust. This was the pseudo-science of eugenics.

The basis of eugenics is the English biologist Charles Darwin's theory of natural selection. Darwin argued that species and individuals are in competition to live and breed and that the individuals and species which are best adapted to their environment will tend to survive and breed with other well-adapted examples; while those which are less well adapted will tend to be less successful at breeding and competing for food and will tend to die out. This ensures that there is a continuous spiral of evolution by which species adapt to the conditions under which they have to live.

When this theory was first advanced in 1859 it was seen as both radical and contradictory of almost all previous teaching on the subject, which hitherto had been largely a religious matter. There was a great deal of evidence that species, including man, did evolve, and a number of scientists and natural philosophers had put forward theories of evolution, but natural selection was such a satisfying and intellectually dazzling solution to the problems posed by the evidence of evolution that it quickly became accepted as the new orthodoxy.

Even as Darwinism was being accepted as the basis of biological evolution, many scientists were taking the theory out of its biological context and adapting it to help them explain other apparently analogous scenarios, the most obvious of which was human society. The apparently rigid class structures of nineteenth-century Europe and America seemed to offer a great deal of scope to a 'social-Darwinist' interpretation, and not only because of its close resemblance to the contemporaneous Marxist analysis. By applying the new Darwinian ideas, it was easily possible to confirm the beliefs and prejudices which underpinned society, and to apply an entirely bogus 'rank-order' or 'value-judgement hierarchy' based on a supposedly universal biological law. Thus, for example, it became supposedly proven that men were more intelligent than women because they had adapted to society by developing a bigger brain. According to the social psychologist Gustave Le Bon, writing in 1879: 'In the most intelligent races, as among the Parisians, there are a large number of women whose brains are closer in size to gorillas than to the most developed male brains. This inferiority is so obvious that no one can contest it for a moment; only its degree is worth discussion,' Not entirely surprisingly, this analysis was also applied to race, and it was easy enough for anthropologists from the major European colonial powers and the USA to conclude that blacks were at the foot of the human hierarchy. Nor were socio-economic groups immune to such differentiation: the rich and well-educated elites clearly possessed a better biological inheritance than the lower – or poorer – classes. The historian Henry Friedlander notes:

> ...the biological sciences of the nineteenth century simply recorded traditional prejudices. Without any evidence, scientists concluded that human differences were hereditary and unalterable, and in doing so, 'they precluded' redemption because they imposed 'the additional burden of intrinsic inferiority upon despised groups'. Science thus showed 'the tenacity of unconscious bias and the surprising malleability of "objective", quantitative data in the interest of a preconceived idea.'

The rediscovery of the genetic theories of the eighteenth-century monk, Gregor Mendel, led scientists to begin to look for individual genes which would control the development of specific, complex body parts and actions and, as a corollary, to come to the conclusion that environment was unable to alter or influence the development of the individual.

Further down this same line there emerged a range of ideas, based on the 'natural selection' explanation of evolution, which proposed that alongside intellectual inferiority came moral inferiority; that the lower forms of human life were also likely to be criminally inclined. Criminality being thus hereditary, punishment would be a waste of time and effort, but 'elimination' might well be a way to eradicate crime from society. One genetic marker of criminality identified at this time by an Italian doctor, Cesare Lombroso, was epilepsy, because '... almost every "born criminal" suffers from epilepsy to some degree'. Nor was he too keen on gypsies, who, he asserted, waste everything they earn on 'drink and ornaments' and could be seen 'barefooted but with bright or lace bedecked' clothes: 'They have the improvidence of the savage and that of the criminal as well.'

The culmination of this strand of social Darwinism was the eugenics movement, which was, during the early decades of the twentieth century, an important talisman of 'progressive', modernist and politically correct thought, not unlike today's environmentalism. The term 'eugenics' had been coined in 1881 by the British scientist Francis Galton, and came to describe the attempt to improve the human race by better breeding, much as one might attempt to breed the perfect poodle or fastest racehorse. In its purest form it was neither especially malignant nor a pseudo-science, in the sense that eugenicists attempted to gather and interpret data as objectively as they could. But, like many modern environmental enthusiasts, the eugenicists of the late-nineteenth and early-twentieth centuries were responding to a set of imperatives and 'crises' that were as much the result of their own inner prejudices and anxieties as they were a response to a real problem affecting society.

The real roots of eugenics were the social, cultural and racial prejudices that affect many, if not all, of us, but expressed in a much more subtle and opaque way. The concern that motivated the pioneers of eugenics was that advances in science, medicine and welfare practice were acting in a way that was counter to natural selection. In effect, that these advances were artificially favouring individuals, groups and races which would, in the natural course of events, be selected for extinction because of their inferior adaptation to modern society. These peoples were, of course, to be found in the groups which had already been identified as the lower orders but the fact that the different societies which took up eugenics tended to pick on different groups as presenting a danger does indicate how subjective these ideas were. In Germany concern centred on fears of being overwhelmed by 'hordes' of Slavs from the east; in the USA worry focused on the arrival of eastern-European Jews and immigrants from southern Europe. By contrast, Scandinavia was concerned that the best examples of 'Nordic types' were emigrating to the USA.

The 'science' of eugenics was taken up with enthusiasm by both the political left and right. Rightists exhibited their traditional prejudices against people from different racial and ethnic backgrounds but the progressive left developed a concern that the working class might be corrupted and tainted by criminal degenerates. The historian Michael Burleigh characterizes it thus:

> Eugenics was simple and it offered a very radical quick-fix solution to very complex problems. It was a totally international movement, not just in

> Europe or indeed North America, but Latin America, China, India, Japan and so forth, and it was cross-party political with just as many enthusiasts on the left as on the right. And the reason that it appealed to the left was that it was against traditional Judaeo-Christian morality at a time when many socialist parties were aggressively atheist, it was apparently progressive, it was apparently ... scientific, and finally it very much appealed, given that most socialist parties represent labour aristocracies – the posher end of the working class – it appealed to their very considerable prejudices against deracinated *lumpen* proletarians. Also, whereas on the right people had an uncomplicated loathing of the poor, on the left loathing was co-mingled with guilt and for many middle-class socialists, particularly doctors and women's groups within socialist parties; eugenics appealed to them as a way of dealing with the unreconstructed poor... A lot of middle-class socialists like Sidney and Beatrice Webb, who were Fabian socialists involved in the founding of the LSE [London School of Economics], had a particular idea of what working-class people should be like and those who didn't conform to that idea they wanted essentially to eradicate and in this case by proposing their sterilization.

Eugenicists firmly believed that Mendelian inheritance governed character traits as well as physical characteristics. Thus it was assumed that, for example, there was a gene which governed an 'attraction to the sea' which would explain why naval careers tended to run in families, and if this was the case, eugenics would have great value in acting as a stabilizing and improving force in society, as Burleigh explains:

> The attraction of the hereditarian analysis and the sorts of solutions that eugenicists were putting forward was that it was very simple and that they didn't then have to deal with very complex interactions between human

> beings and their environments, they didn't have to think too hard about ... complex social and economic questions and eugenics provided a very radical, apparent quick fix to all these things; you could cut through the complexity. [For example] they're assuming that alcoholism is hereditary, which of course it might not be ... they're constantly cutting through what are probably the product of the interaction between our genes and the environment, and saying this is all to do with our genes and that we simply have to sort of mess around with that by sterilizing people or whatever, or isolating them and we have solved the problem...

The result was eugenics as an academic science. The first university chair in the subject was established at University College London in 1909, but many others followed as the subject was taken up with enthusiasm around the world. In essence, the goals of academic eugenics were twofold: to investigate and identify inherited traits, diseases and behaviour, and to codify the groups and individuals who bore them on a scale of human worth; and to apply the knowledge thus gained in order to bring about biological solutions to social problems.

The problem-solving potential that eugenics seemed to offer was attractive to politicians as well as scientists. Breeding out unhealthy groups promised to save money and rid society of an apparent burden, and policy makers took to it with alacrity. As early as 1899 the state of Indiana sought to enact legislation for the compulsory sterilization of the mentally handicapped – inevitably regarded by eugenicists as depraved and immoral and certain to give birth to degenerate children – and was eventually followed in this by more than thirty other American states. This eugenic practice was so accepted by society that when the constitutional validity of the Commonwealth of Virginia's sterilization legislation was challenged in the US Supreme Court, the majority opinion upheld the law with considerable vigour: 'It is better for all the world, if instead of waiting to execute degenerate offspring for crime, or to let them starve for their imbecility, society can prevent those who are manifestly unfit from continuing their kind. The principle that sustains compulsory vaccination is broad enough to cover cutting the Fallopian tubes.'

Similar policies, for similar reasons, were applied in Scandinavia (where they remained in place well into the 1970s).

The growth of eugenics in Germany followed a similar pattern to those of Britain and the USA, although the organization and traditions of German academia ensured that it was much more centralized, and where eugenic groups were scattered about at the local level in Britain and the USA, in Germany they were under a single umbrella, the German Society for Race Hygiene.

Ploetz and Schallmeyer, the founders of the German eugenics movement before the First World War, were in favour of 'positive eugenics'. They were worried by the invasions of 'Slavic hordes' and the operation of social welfare among the lower orders, but they did not believe that German society would support a policy of compulsory sterilization. Instead they advocated measures which would, in effect, improve the 'breeding stock' of the German people. However, they differed on what this might mean and their disagreement prefigured a divergence in German eugenics: Ploetz believed that the best examples of 'Germanhood' would be drawn from the Nordic or Aryan 'race'; whereas Schallmeyer was in favour of more objective measurements of IQ, education, wealth and so forth. The result of this disagreement was a split among German eugenicists between 'Aryan supremacists' and the rest which was to last until the beginning of the Third Reich.

The intellectual underpinning of German eugenics was developed during the period just before the First World War. One of the most influential texts from this time was a study conducted in 1908 and published in 1913 by the anthropologist Eugen Fischer of the offspring of mixed marriages between Dutch-speaking Boers and native Hottentot women in the German colony of South West Africa (now Namibia). Fischer's conclusion was that the offspring of such mixed marriages were invariably of an inferior type: 'We still do not know a great deal about the mixing of races. But we certainly do know this: without exception, every European people that has accepted the blood of inferior races – and only romantics can deny that Negroes, Hottentots and many others are inferior – has paid for its acceptance of inferior elements with spiritual and cultural degeneration'. The upshot of this and other work was the banning of mixed marriages in Germany's African colonies and consideration was also given to extending similar measures to Germany itself. Dr Wolfgang Eckhardt has studied the involvement of German doctors in the eugenics movement:

> The whole ideology was very idealistic and very modern at the time. People thought that it could be possible to create a better nation by [positive eugenics] and they thought themselves to be at the front of a very modern medical movement with that. Genetic diseases had been discovered very recently and people were thinking about measures to prevent them. That was for the first time that people ... were thinking about how to prevent diseases, besides bacteriology, where this thinking was coming up at the same time.
>
> There is a second discipline, quite important for this aspect of modernity of medicine: at that time people were not thinking of

aggressive sterilization measures or acts even of aggressive forced sterilization – sterilization laws – they were thinking of the juridical aspects, they were thinking of laws [to ban mixed marriages and so forth]. But nobody then was planning to force people to be sterilized or, even worse than that, to be killed because of their genetic deformities.

In Germany – well, it differs probably from country to country and from cultural situation to cultural situation – it was ... combined with the idea of Aryanism, what they wanted to create was not only a healthy person and a healthy nation; they wanted to create an Aryan nation. That means the white Germanic race should be supported by those measurements, that was the idea which fits very, very well into German imperialism of that time [and] which was one of the roots that led into the First World War. Paradoxically this idea collapsed with the war itself and after the war they seemed to have good reasons then to think about racial hygiene again because they had this degenerative, as they thought, aspect and effects of the war: the best being killed, the worst remaining, surviving and then producing children again and again. They had diseases: alcoholism and tuberculosis increasing, things like that; which were thought of as important aspects to produce ... genetically deformed persons. So this was a kind of revival: a second spring of racial hygiene; and from that one – from the early twenties – on the way, the pathway to what happened after '33 was prescribed quite clearly.

Thus, almost by chance, the early Nazism of the dispossessed lower middle class and 'betrayed' ex-servicemen came into contact with, and began to appropriate, the apparently respectable science of eugenics. There is no really convincing account of how Adolf Hitler became a virulent and obsessive anti-Semite but his most recent biographer, Ian Kershaw, in *Hitler 1889–1936: Hubris*, sums up the most likely course thus:

> It seems more likely that Hitler, as he later claimed, indeed came to hate Jews during his time in Vienna. But, probably, at this time it was still little more than a rationalization of his personal circumstances rather than a thought-out 'world view'. It was a personalized hatred – blaming the Jews for all the ills that befell him in a city that he associated with personal misery. But any expression of this hatred that he had internalized did not stand out to those around him where anti-Semitic vitriol was so normal. And paradoxically, as long as he *needed* Jews to help him earn what classed as a living, he kept quiet about his true views and perhaps even on occasion, as Hanisch indicates, insincerely made remarks which could be taken, if mistakenly, as complimentary to Jewish culture. Only later, if this line of argument is followed, did he rationalize his visceral hatred into the fully-fledged 'world view', with anti-Semitism as its core, that congealed in the early 1920s. The formation of the ideological anti-Semite had to wait until a further crucial phase in Hitler's development, ranging from the end of the war to his political awakening in Munich in 1919.

But eugenics was to provide Hitler with an apparently solid and respectable scientific support for his hatred. On the evening of 8 November 1923 a force of Nazi Party supporters, with Adolf Hitler at their head, surrounded the Bürgerbräukeller in Munich, where a political meeting was taking place. Waving a pistol and shouting noisily, Hitler entered the room, jumped on to a chair and fired his weapon into the ceiling to gain attention. With all eyes now focused on him, he declared that a national revolution had started and that the hall was surrounded by 600 armed men. After ushering the Bavarian Prime Minister, police chief and army chief into a nearby room, he left it to his friend Hermann Göring to quieten the mob. Göring swiftly took control; stay calm and remain in your places, he told them. 'You've got your beer.'

But after the first few tumultuous hours Hitler's putsch against the national government began to go badly wrong. Promises extracted from the Bavarian leaders at pistol point were reneged on as soon as the hostages had extricated themselves from Nazi custody, and as Hitler and his followers floundered about, attempting to grab control of the levers of state power in Bavaria, the state government, police and army organized to strike back. By the next morning the initiative was firmly back in the hands of the authorities and as a last despairing effort, Hitler and the putschists organized a march through Munich in the hope of attracting popular support for their cause. As they reached the central Odeonsplatz they were confronted by a police cordon and firing broke out, leaving fourteen putschists dead and many more wounded. Hitler, with a severely dislocated shoulder, was bundled into a car and driven away from the scene by a supporter. Two days later he was arrested at the home of a friend near Munich and imprisoned in the fortress at Landsberg.

The 'Beer-Hall putsch' was an essentially opportunistic act by Hitler and the Nazis, provoked by the political crisis which had developed in Germany as a result of the French occupation of the industrial Ruhr, itself an act of retaliation for Germany's inability to pay war reparations. Germany was essentially bankrupt and hyperinflation made the Reichsmark worthless: on the day of the putsch the *Völkischer Beobachter*, the Nazi Party newspaper, was on sale for 5,000 million marks; a week later one US dollar would buy 4.2 trillion marks. Although the putsch was a farcical failure, it did have the effect for Hitler of elevating him from being a purely local phenomenon in Bavaria to the national stage: it also gave him slightly more than a year in prison in which to study and reflect: 'Landsberg was a university paid for by the state,' he later told his lawyer, Hans Frank.

One of Hitler's supporters during his imprisonment was a Munich publisher, Julius Friedrich Lehmann, who specialized in publishing both medical textbooks and nationalist tracts. In 1921 he had published the *Grundrisse der menschlichen Erblehre und Rassenhygiene* (Outline of Human Genetics and Race Hygiene) in two volumes by Erwin Baur, Eugen Fischer (of the study of Hottentots) and Fritz Lenz, which was to become the bible of German eugenics. In 1923 Lehmann issued a second edition, and according to the German biologist and historian Benno Müller-Hill: '...arranged that [Hitler] got a copy in the Landsberg jail, where he spent a year and ... he clearly read it because there's one chapter in *Mein Kampf* which clearly ... shows that he read the thing and gives quasi-quotes from the textbook.'

The importance of this wasn't so much in the ideas that Baur, Fischer and Lenz were putting forward, but in the way that they gave a scientific legitimacy to Hitler's anti-Semitism, as Müller-Hill argues:

> The situation is ... that Hitler's basic drive was anti-Semitism, but of course in this respect the scientists, the biologists were very useful. They were not particularly anti-Semitic, but they could give them reasons for anti-Semitism. So ... there came a cooperation between the scientists who wanted to have all their science going and genetics going and who had to accept the genetic reasons for anti-Semitism and ... so the scientists had to accept anti-Semitism and Hitler had to accept the scientists.

> ...Even before the Nuremberg laws [discriminating against Jews] came, for example, Fischer, the top human geneticist in Germany, asked that a law like this should come. And what more do you want? You want to have anti-Semitic laws from your gut feeling and then come scientists and say that it's scientifically necessary that this is going to come. Then you must be right after all.

Thus, throughout the period of struggle in the 1920s and early 1930s, before the Nazis took power, their racial assumptions – which nowadays constitute perhaps the most offensive aspects of their political programme – could in fact be presented as scientific fact, backed up by and based on research by scholars at some of Germany's most prestigious scientific institutions.

Less than ten years after the farce of the Munich putsch, on 30 January 1933, Hitler became Chancellor of the German Reich, 'jobbed into office by a backstairs intrigue', as one commentator described it. The non-Nazi right-wing nationalists who had made this possible – Hitler controlled the largest party in the Reichstag, the German parliament, but had nothing like an overall majority – thought that they had 'hired' him: they were wrong. In the immediate aftermath of his appointment he was anxious to call a general election in order to consolidate his grip on power. This was held on 5 March 1933 and raised the Nazi Party's share of the vote to 43.9 per cent, helped in part by an upsurge in support caused by the public's anxiety over an arson attack on the Reichstag building by a deranged Dutch communist, Marinus van der Lubbe. With their nationalist coalition partners gaining 8 per cent, the Nazis now had a bare majority to see through their programme.

The key to the Nazis' ability to put their programme into action was the Enabling Act of 23 March 1933, by which the Reichstag abrogated its legislative powers in favour of the Reich Government under Hitler. This was passed by the Reichstag with opposition only from the deputies of the Social Democratic Party (the Communists had already been effectively banned in the wake of the Reichstag fire), thus leaving the way clear for Hitler's government to take summary legislative action for the 'protection of the state'.

Even before any official moves had been made to put the Nazis' anti-Semitic hatred into practice, the first few weeks after their assumption of power saw hundreds of attacks on Jews and their property, carried out by radical thugs fired up by years of anti-Semitic propaganda. Hitler probably sympathized with this, but the attacks received wide publicity outside Germany, and led, particularly in the USA, for calls to boycott German goods and trade until the attacks could be stopped. Hitler's reaction to this was to instigate a counter-boycott of Jewish businesses in Germany, and Julius Streicher, a notorious Jew-baiter and editor of the pornographic propaganda rag *Der Stürmer*, was given the task of organizing it for 1 April 1933.

This episode gave momentum for the first pieces of anti-Semitic legislation to be passed by the German government. On 7 April the Law for the Restoration of the Professional Civil Service was proclaimed, excluding Jews and political opponents from the civil service. This was followed in the same month by three further laws which restricted the entry of Jews into the legal profession, prevented Jewish doctors treating patients under the national health insurance scheme and limited the number of Jewish schoolchildren permitted in schools.

The effect of the 'civil service law' was wide-ranging. Aside from the civil service itself, from national down to local level, a broad sample of institutions within the German state were ultimately funded or controlled by the state itself, including universities and hospitals. Consequently, not only civil servants, but also academics and doctors, found themselves thrown out of work without compensation or redress. Otto Frisch, a young Jewish physicist, was working in Hamburg in the research group of Otto Stern, who also came from a Jewish background:

> It soon became clear that [Hitler's] anti-Semitism was not just talk, and when his racial laws were passed, Stern was quite shocked to find that I was of Jewish origin, just as he was himself and another two of his four collaborators. He would have to leave and the three of us as well, only one of his outfit – Friedrich Knauer – being Aryan and able to remain in a university post. Actually the university in Hamburg – with the traditions of a free Hansa city – was very reluctant to put the racial laws into effect, and I wasn't sacked until several months after the other universities had toed the line. At first I still hoped that I might be able to take up a fellowship which the Rockefeller Foundation had awarded me (at Stern's instigation) to enable me to work in Rome for a year, an opportunity I had very much looked forward to... But the fellowship was conditional on my having a permanent post to go back to; when Hitler's laws came into effect the Rockefeller Foundation regretfully informed me that under those circumstances they could not offer me the fellowship any longer...
>
> Disturbing rumours were rife. Some of my Jewish friends had warned me not to be out at night because Jews had been beaten up in the dark.

It was not only young and unknown scientists who were at risk. Albert Einstein, head of the Kaiser Wilhelm Institute for

Theoretical Physics in Berlin, discovered in America that his home had been raided by Brownshirts and decided not to return. Disgracefully, there was little opposition from scientists to the measures. However, Max Planck, the overall head of the Kaiser Wilhelm Gesellschaft, called on Hitler and raised, with some bravery, the issue of the Jewish scientists:

> After Hitler seized power, it was my duty, as President of the KWG, to pay my respects to the Führer. I thought I should put in a good word for my Jewish colleague, Fritz Haber. I said that the last war would have been lost from the outset had it not been for his techniques for producing ammonia by fixing nitrogen from the air. Hitler answered in these words: 'I have nothing against Jews as such. But the Jews are all Communists, and the Communists are my enemies. It is against them that I fight.' I commented that, even so, there do exist different types of Jews, some of value to humanity and some of no value, and that, among the former, there were long established families who had absorbed the best of German culture. I said that we should make a distinction between them. He replied, 'That is not correct. A Jew is a Jew; all the Jews stick together like burrs. Where there is one Jew, all kinds of other Jews gather. The Jews themselves should have established those distinctions. They did not do so and therefore I must proceed against all of them equally.' I then remarked that we were inflicting damage on ourselves by forcing those Jews whose talents we needed to emigrate and that their talents would now be used for the benefit of foreigners. This he did not accept at all and held forth at great length about quite general matters, ending up by saying: 'It is said that I suffer on occasion from weak nerves. That is a slander. I have nerves of steel.' With that, he slapped his knee with great force, spoke more and more rapidly and began to shake with such uncontrollable rage that there was nothing I could do but keep silent and take my leave as soon as I decently could.

The Kaiser Wilhelm Gesellschaft, although in theory independent, was soon informed that its fifty-four 'non-Aryan' employees would have to go as well.

In the field of medicine the effects were similar. A fairly large number of German doctors at this time were of Jewish origin, because, as Burleigh explains: 'It's rather like law, in that it's a secure profession and I suppose if you were a recent immigrant it's one of the things you would do rather than say philosophy or history.' As a result:

> ...one of the main hard-core constituencies of anti-Semitism in Germany and Austria and probably elsewhere were university students. And in the case of medical students they felt threatened by an influx into their courses, and their lecture halls, of East European Jews who wished to study medicine as a relatively secure career... And this they carried over into their adult life, and then would alight upon what they alleged to be the large number of Jewish doctors in particular cities or in particular branches of medicine.

Thus, for many doctors, the new state-sponsored discrimination was to be welcomed rather than condemned, as Wolfgang Eckhardt describes:

> ...for the larger part of the German physicians, especially for the younger ones which had been joining the Nazi movement very early, they didn't have any scruple when it came to expelling their Jewish colleagues. They were quite eager to fill the posts they left: it was a good job. After only a few years more than 50 per cent of the Jewish posts that were left in Berlin by expelling those colleagues were occupied again by German – by Aryan – physicians. That was the purpose ... I think it was just an economic aspect which made it quite important for the physicians.

It was now that one of the most remarkable features of Nazism began to become apparent, the sense in which it has been described as 'politics as applied biology'. The Nazis genuinely believed that it would be possible to resolve social and political problems by biological means. The first manifestation of this belief, and the model for all German eugenic legislation thereafter, was the 'sterilization law' of July 1933. Eugenicists had been advocating sterilization of 'inferior and degenerate' types for some years and proposals for voluntary sterilization had made it as far as state legislatures in Germany. The new law was different in that it introduced compulsion: 'Any person suffering from a hereditary disease can be sterilized if medical knowledge indicates that his offspring will suffer from severe hereditary physical or mental damage', ran the opening preamble. The following diseases were classified as hereditary within the meaning of the law:

1. Congenital feeble-mindedness
2. Schizophrenia
3. Manic depression
4. Hereditary epilepsy
5. Huntington's Chorea
6. Hereditary blindness
7. Hereditary deafness
8. Severe hereditary physical deformity
9. Severe alcoholism

The structure for enforcement was straightforward. If the handicapped person did not voluntarily apply for sterilization, it could be sought by health service doctors and directors of hospitals, homes and prisons. A system of Hereditary Health Courts was instituted, consisting of three members: a judge and two doctors; with an appeal court at a more senior level, although structured in the same way.

In the first year in which the sterilization law was in operation, some 388,400 people were denounced to the Hereditary Health Courts, 75 per cent by their doctors. This was too much work for the courts and only just over 80,000 of these came before them, of which some 62,000 were made the subject of a sterilization order. Of these, a little under half were sterilized (usually by vasectomy for men and fallopian ligations for women) as the result of a lack of capacity in the hospitals, and the system does not seem to have caught up before the outbreak of war in 1939. Despite the increase in workload, the reasons for the German medical profession's enthusiasm for the sterilization programme are not hard to find, according to Michael Burleigh:

> It gave them an enhanced range of tasks for which they were paid, and the introduction of eugenics meant that there was a new, large bureaucracy involved in implementing it and doctors benefited financially from doing that. The biggest losers, of course, were patients, because they were being neglected because the doctors were so heavily engaged in filling out eugenic information forms and acting as expert witnesses to the new range of Hereditary Health Courts.
>
> ...On the basis of the flimsiest amount of actual scientific information these people made very large claims for what science could achieve and participated in what you might call a gigantic act of faith which was, of course, very beneficial to them because the government rewarded them for the large claims that they were making.

> It's customarily assumed that doctors dispose of a vast fund of knowledge and that they're essentially idealistic in their motivation and, as this case shows, many of them clearly weren't ... The doctors had moved from being concerned about sick individuals to essentially being concerned about the genetic health of the collective, which, as they perceived it, elevated their function. And they were fully prepared to cut all corners to implement those policies, so checking the scientific basis of the policies that they were being asked to implement went out of the window, and ethical concerns went out of the window, and then the bureaucratic burden that they were subjected to positively encouraged them to cut corners.

Wolfgang Eckhardt agrees with this analysis:

> The physicians were encouraged to see themselves as physicians not only for the individual but, very much more strongly, for the race. This is the major change in 1933: that the very individualistic medicine – the individual to care for the individual, for the patient – changed into a care, a medicine, for the race: for the Volk, for Reich and Führer, of course. So the single person was not worth anything any longer but the nation was worth everything, and this is, I think, a very clear-cut difference in medical thinking before 1933 and after 1933; and it's a very close link to the Nazi takeover in early 1933. And would it make the doctors feel more powerful and important? Instead of being just that man who gives you aspirin you suddenly become this person who has a really big powerful role, don't you? It changes your self-image.

Despite the learned tracts which had acted as the basis for the sterilization law, and which underpinned Nazi 'politics as applied biology', the scientific basis for what the German doctors was doing was, according to Eckhardt, deeply flawed:

> The sterilization measures could have never been successful: seen from a biological point of view, they are useless, absolutely nonsense, because they do not calculate, or people didn't calculate, spontaneous mutations, hereditary deformations by poisons, by outer [environmental] poisons probably and things like that. So it was a measure that would have never led to success – to real success – even if they had practised it very harshly, even more harshly than they did.
>
> So it was senseless from the beginning, but it was an important part and a kind of sign for the biological takeover that National Socialism was planning: it was not only a political takeover but also a biological takeover. And they had people sterilized, more than 350,000 between the years 1933 and 1945, so it was also a kind of discrimination against people that wouldn't fit into the picture of a Nazi society. So everything else, looking bad, looking ill, looking strange, had to disappear: that was their biological plan for a biological revolution, for a biological takeover

Although the number of people who fell into several of the categories was – to some extent – finite or at least limited by diagnosis, in the case of 'congenital feeble-mindedness' and alcoholism, application of the law could be made only on the basis of subjective social judgements. As in the USA, intelligence tests were introduced which would supposedly give a conclusive result but which were, in reality, tests of acquired learning open to subjective interpretation. Henry Friedlander describes the case of Erwin Amman, a twenty-one-year-old from the Tyrol, who, despite answering most of the questions correctly and promptly, was sterilized because he displayed 'feeble-minded appearance and behaviour'. On the other hand, in another case, a health service doctor in Gera denounced a twenty-five-year-old factory worker for feeble-mindedness, but had the case rejected by the court. The doctor appealed against this decision and further investigations were made, but the appeal court upheld the original decision, having concluded that, although not terribly bright, the individual was perfectly aware of what was happening in the world and capable of looking after himself.

A split began to emerge in the late 1930s between the health professionals who were enthusiastically enforcing sterilization and certain Nazi party officials who had some qualms about it. If you can find individuals with congenital diseases, then other members of their families will be affected as well. The logic of this argument was not lost on public-health officials, who began to trawl through private medical records and family trees; or to chase after individuals convicted of petty offences which might indicate a supposedly congenital condition like alcoholism; to the evident annoyance of some senior Nazis. Burleigh quotes the deputy Reich Doctors' leader in 1936:

> When a peasant boy from Masuria who has hardly had any experience of schooling because he always had to work in the fields, comes to Berlin and joins some party formation or other, and then commits some stupidity while drunk, this is soon followed by an application to have him examined with a view to his eventual sterilization. Then there is the famous questionnaire, e.g. 'When was Columbus born?' and the boy answers no to everything saying 'I don't know anything about that', because it is quite possible that he has never had the chance to learn these things. A doctor who has examined him only once certainly cannot, on this basis, come to a final verdict that the boy is of lesser value, because perhaps his abilities have never been able to come to fruition.

In essence, this Nazi doctor encapsulates the problem with eugenics: it was then and remains a blunt instrument attempting to make use of an immensely complex science, genetics, which is still poorly understood (and in the 1930s was barely understood at all). The similarity with modern environmentalism is striking:

> When I look at the situation in 1933, what the scientists are lacking really to enable them to develop a feasible eugenic programme is the diagnostic instrumentarium: they assume that mental illnesses and mental handicap are to a very great extent conditioned by the hereditary factor, but they cannot prove in individual cases how that happens.
>
> Nevertheless, they already want to start with sterilizations on a large scale in order to act preventively. I always say: that is similar to some ecologists wanting a speed limit on the autobahn, because they assume that car exhaust fumes are destroying the forest, but they can't exactly prove it but they say, 'We must act preventively, otherwise by the time we can provide the proof there will be no forest left.'
>
> At the beginning of the thirties they argued in a similar way in the field of eugenics. So what they were lacking was the diagnostic instrumentarium. They couldn't really prove that schizophrenia was due to heredity, or that a form of feeble-mindedness was inherited and not acquired.

A joke popular in Germany in the 1930s summed up the paradox of the senior Nazis' enthusiasm for eugenics and racial classification in its definition of an Aryan: 'He must be blond like Hitler, thin like Göring, handsome like Goebbels, virile like Röhm – and be called Rosenberg.' Discriminatory laws based on eugenic justifications continued to be enacted throughout the decade. The sterilization law was eventually complemented by laws which confined anti-social individuals (*Asozialen*) to state hospitals or asylums, imposed protective custody on habitual criminals, restricted the right of gypsies to travel and trade (automatically classifying them, on the basis of their race, as anti-social criminals) and, in the Law for the Protection of German Blood and German Honour, deprived Jews and other 'alien' races of their German citizenship:

NUREMBERG LAW FOR THE PROTECTION OF GERMAN BLOOD AND GERMAN HONOUR, SEPTEMBER 15, 1935

Moved by the understanding that purity of the German Blood is the essential condition for the continued existence of the German people, and inspired by the inflexible determination to ensure the existence of the German Nation for all time, the Reichstag has unanimously adopted the following Law, which is promulgated herewith:

Article 1.

1. Marriages between Jews and subjects of the state of German or related blood are forbidden. Marriages nevertheless concluded are invalid, even if concluded abroad to circumvent this law.

2. Annulment proceedings can be initiated only by the State Prosecutor.

Article 2.

Extramarital intercourse between Jews and subjects of the state of German or related blood is forbidden.

Article 3.

Jews may not employ in their households female subjects of the state of German or related blood who are under forty-five years old.

Article 4.

Jews are forbidden to fly the Reich or National flag or to display the Reich colours. They are, on the other hand, permitted to display the Jewish colours. The exercise of this right is protected by the State.

Article 5.

1. Any person who violates the prohibition under §1 will be punished by a prison sentence with hard labour.

2. A male who violates the prohibition under §2 will be punished with a prison sentence with or without hard labour.

3. Any person violating the provisions under §3 or §4 will be punished with a prison sentence of up to one year and a fine, or with one or the other of these penalties.

Article 6.

The Reich Minister of the Interior, in coordination with the Deputy of the Führer and the Reich Minister of Justice, will issue the Legal and Administrative regulations required to implement and complete this Law.

Article 7.

The Law takes effect on the day following promulgations except for §3, which goes into force on January 1, 1936.

Nuremberg, September 15, 1935 at the Reich Party Congress of Freedom

The Führer and Reich Chancellor Adolf Hitler

The Reich Minister of the Interior Frick

The Reich Minister of Justice Dr Gürtner

The Deputy of the Führer R. Heß

Clearly, however, for a truly committed eugenicist, sterilization and legal discrimination can be no more than a halfway house towards the ultimate goal of a genetically and racially pure society, and the Third Reich was to provide eugenicists with the opportunity to go many steps beyond this in their quest to solve political problems with biology.

But genetic purity was not the only goal of Nazi medical biology:

> In many ways, my sheepdog Blondi is a vegetarian. There are lots of herbs which she eats with obvious pleasure, and it is interesting to see how she turns to them if her stomach is out of order. It is astonishing to see how wise animals are, and how well they know what is good for them...
>
> Whenever I have to make a speech of great importance I am always soaking wet at the end, and I find I have lost four or six pounds in weight. And in Bavaria, where, in addition to my usual mineral water, local custom insists that I drink two or three bottles of beer, I lose as much as eight pounds. This loss of weight is not, I think, injurious to health. The only thing that always worried me was the fact that my only uniform was a blue one, and it invariably stained my underclothes!
>
> When I gave up eating meat, I immediately began to perspire much less, and within a fortnight to perspire hardly at all. My thirst too, decreased considerably, and an occasional sip of water was all I required. Vegetarian diet, therefore, has some obvious advantages. I shall be interested to see whether my dog eventually becomes a complete and confirmed vegetarian.

The mild-mannered, animal-loving vegetarian responsible for the above musings was, of course, Adolf Hitler. In addition to his dislike of meat, he had little time for smoking either:

I made the acquaintance in Bayreuth of a businessman, a certain Möckel, who invited me to visit him in Nuremberg. There was a notice above his door: 'Smokers not admitted.' For my part, I have no notice above my door, but smokers aren't admitted.

Some time ago, I asked Göring if he really thought it a good idea to be photographed with a pipe in his mouth. And I added, 'What would you think of a sculptor who immortalized you with a cigar between your teeth?'

It's entirely false to suppose that the soldier wouldn't endure life at the front if he were deprived of tobacco. It's a mistake to be written on the debit side of the High Command, that from the beginning of the war it allotted the soldier a daily ration of cigarettes. Of course, there's no question now of going into reverse. But as soon as peace has returned, I shall abolish the ration. We can make better use of our foreign currency than squandering it on imports of poison.

I shall start the necessary re-education with the young. I'll tell them: 'Don't follow the example of your elders.'

I experienced such poverty in Vienna. I spent long months without ever having the smallest hot meal. I lived on milk and dry bread. But I spent thirty kreuzers a day on my cigarettes. I smoked between twenty-five and forty of them a day. Well, at that time a kreuzer meant more to me than 10,000 marks do today. One day I reflected that with five kreuzers I could buy some butter to put on my bread. I threw my cigarettes into the Danube, and since that day I've never smoked again.

I'm convinced that, if I had continued to be a smoker, I'd not have held out against the life of incessant worry that has for so long been mine. Perhaps it's to this insignificant detail that the German people owes my having been spared to them.

So many men whom I've known have died of excessive tobacco. My father, first of all.

Hitler the New Labour-style health nanny is an unfamiliar concept to most people, but Nazism, in its drive to turn politics into applied science, was as keen to eliminate environmental risks to the health of the national collective as it was to drive out hereditary contamination. Under the Third Reich, doctors were among the first to recognize and identify many of the factors that caused cancer, including industrial smog, asbestos, radiation and tobacco; they recognized and extolled the importance of a healthy diet, surprisingly similar to those advocated by public-health lobbyists today; and they campaigned to reduce smoking and drinking.

A healthy, potent state requires a healthy, potent population. In the Third Reich, the welfare of the individual citizen was subordinated to the needs of the state, and the state required 'tough, lean, high-performing man machines'. Maintaining good health, therefore, became a key task for the dutiful citizen, and much of the Nazis' efforts in this direction focused on the national diet.

In the first half of the twentieth century post-mortem examinations demonstrated that gastric malignancy accounted for by far the largest proportion of cancer deaths in Germany, Switzerland and the USA. Part of the reason for this stemmed from the poor quality of the food that was eaten then. Long-term storage methods were primitive and often ineffective: salting, smoking and fermenting were common ways of preserving food, but these didn't necessarily get rid of many of the microbes which were present, while the preservation methods themselves frequently added dangerous and potentially carcinogenic agents. The Nazis were obsessed by racial purity, but these obsessions carried over into their views on diet as well.

Studies showed that, in the early part of the nineteenth

century, the average German ploughed his way through an annual 14 kilograms of meat and 250 kilograms of grains. By the mid-1930s this picture had changed completely, and they were consuming 56 kilograms of meat and only 86 kilograms of grain, while consumption of sugar had increased sixfold from 4 to 24 kilograms. Franz Wirz, a member of the Nazi Party's Public Health Committee, ascribed to these changes an increase in tooth decay, together with a clutch of ailments, including infertility, stomach tumours and nervous complaints. Dietetic science at this time was concentrating largely on the energy content of food to calculate its nutritional value, but Wirz and others argued that the National Socialist diet needed to be much more natural, containing fewer preservatives, colourings and other additives, and not overcooked, to reduce loss of valuable vitamins and minerals. The Nazi diet would therefore be not only healthier, but also much more economically viable.

Vegetarianism was a personal enthusiasm of the Führer, but also of several other leading Nazis, including the Reichsführer-SS, Heinrich Himmler. This was partly the result of their concerns about cancer, but also because of a strange sentimentality about animals. Hitler would wax lyrical about Foxl (Foxy), a fox terrier he had befriended in the trenches of the Western Front during the First World War, and was probably emotionally closer to his German shepherd dog Blondi than he was to any other member of his entourage (although this didn't stop him from having a suicide pill tested on the dog in the closing days of the war). In September 1933 Hermann Göring outlawed animal vivisection in the *Land* of Prussia, of which he was Minister-President, despite having by then conspired in the opening of the brutal concentration camps at Dachau, Sachsenhausen, Columbia Haus and elsewhere. In fact, although Hitler would hold forth at length about the scientific basis of vegetarianism – he believed it to be the original human diet – he had originally taken it up in order to lose weight he had gained while incarcerated in Landsberg prison.

Alcohol was also frowned upon in Nazi circles. Hitler himself was not a strict teetotaller, but rarely drank anything stronger than mineral water and was followed in this by Himmler, among others. Alcohol was blamed for causing cancers among those employed in the licensed trade, as well as for 'deadening the nerves, weakening the spirit' and even inducing genetic damage. Moreover, the Nazis were early opponents of drink-driving, and in 1937 Himmler, as Reich Interior Minister and Chief of the German Police, wrote to all of Germany's licensed drivers warning against its perils.

But to return to cancers, perhaps the most surprising and neglected research into the subject carried out under the Third Reich was that linking smoking and lung cancer.

In his admirable book *The Nazi War on Cancer*, Robert N Proctor suggests that the link between tobacco and lung cancer was originally 'established' in Nazi Germany. This is not quite correct: in the eighteenth and nineteenth centuries doctors in Britain, France and Germany had linked cancers of the mouth, nasal passages and lips to pipe-smoking, while lung cancer itself remained an extreme rarity. But in the 1920s and 1930s there was a sudden huge and inexplicable rise in cases of lung cancer.

A number of theories to explain this were tried out: the use of the internal combustion engine had risen hugely, power tools and modern production processes were creating clouds of potentially carcinogenic dust and smoke, and the asphalting of roads had increased production of noxious tars at least fortyfold.

But it wasn't until 1929 that a German physician, Fritz Lickint, published a paper giving evidence of a statistical link showing that lung cancer patients were especially likely to be smokers. What had happened was the arrival of the cigarette: cigarettes used milder tobacco and, as a result, smokers were able to draw the smoke further into their lungs than pipe or cigar smokers could. The First World War had massively boosted rates of cigarette smoking, as soldiers and others had used cigarettes to 'calm' their nerves.

Further research by Lickint during the 1930s led him to claim that smoking was in reality the cause of cancers of the mouth, jaw, throat, windpipe and lungs; of hardening of the arteries; infant mortalities; halitosis and many other diseases. He also pointed out the dangers of passive smoking and drew attention to the addictive qualities of nicotine.

Lickint wasn't a Nazi – he had been a supporter of the Social Democrats before 1933 – but his work fitted in well with the Third Reich's quest for purity and 'medical moralism' and was taken up with considerable enthusiasm. Sir Richard Doll, who pioneered post-war research into the link between smoking and cancer, was somewhat sceptical of the Nazis' motivation:

> Well, the Nazi government from the time it first came in took steps to discourage smoking because Hitler was very much against it himself, but the reasons were reasons which wouldn't have appealed to people in this country, and I may say it didn't appeal to the German public, as far as I can make out. Very few people paid attention to the discouragement, but the reasons were firstly that it damaged the hereditary material of the German race, that it reduced female fertility and that the addiction to tobacco would reduce people's willingness to carry out the Führer's

> orders. So it was not really reasons for discouraging smoking that would have appealed to Englishmen, it wasn't on the basis of the diseases that smoking caused.

Lickint was followed by a study in 1939 by a young doctor, Franz Müller, about whom very little is known, who wrote a dissertation described by Proctor as 'the world's first controlled epidemiological study of the tobacco-lung cancer relationship'. Müller studied ninety-six cases of fatal lung cancer and compared them with eighty-six healthy 'controls', eventually reaching the conclusion that 'the extraordinary rise in tobacco use' was 'the single most important cause of the rising incidence of lung cancer'. This was a strong conclusion; far stronger than that reached by Richard Doll in his paper of 1950, in which he was only prepared to cite smoking as '... an important factor, in the production of carcinoma in the lung'.

In fact Doll had studied in Germany in the 1930s:

> Germany at that time was the leading scientific country and had been for some years, particularly in subjects like chemistry and medicine. If you wanted to be a medical research worker you had to go and study after you qualified in Germany, just like nowadays you nearly always have to go and study in America for a short time... I started learning German and as a student I went for a fortnight on a trip that was organized by one of the senior people in our hospital. For one fortnight one year we went to Amsterdam and the next year we went to Frankfurt and that was a very interesting experience because it was after the Nazis had come to power.
>
> I went, of course, every day to lectures in the morning, did other things in the evening and I went to one lecture given by an absolutely delightful man by the name of Wohlhart, internationally known for his knowledge of

renal disease, and he was a splendid character and as I understand it he stood out against the Nazis the whole time and was really considered too important internationally for them to touch ... though I believe he lost his senior position. But I went to another which was a very different matter, given by a radiotherapist who was an active Nazi and he gave us a lecture on radiotherapy. We were told beforehand that we should all stand up and give the Hitler salute when he came in: well you can imagine what twenty-five young Englishmen did when they were told that.

We didn't do it but he gave us a lecture and he showed a slide which has stuck in my memory ever since because I think it was just about the foulest thing I can remember having seen. He showed a slide of X-rays damaging cancer cells, and the X-rays were illustrated by having Nazi Stormtroopers sort of seated on the ray and the cancer cells were Jews ... well, it's certainly the most foul thing I ever heard in a lecture.

Doll was aware of German epidemiological studies on cancer from an early stage:

I know just about some of the research which was done in the Nazi period ... concerning the subject I was especially interested in, namely the effect of smoking. And not surprisingly one found that, in fact, in Germany they had been more concerned, more interested, in smoking for a long time, before the Nazi period, than we had been in England. There were two papers on lung cancer and smoking brought out in the Nazi period, one by a man called Müller and another later on during the war by two people, Schairer and Schörniger. They both suggested a relationship between lung cancer and smoking.

Neither of them were what one could have called good epidemiological papers: much better epidemiology was being done in England earlier than that; but it was data which we hadn't had in England and was suggestive. Although, as I say, nowadays we would call it bad epidemiology.

German scientists at this time also noted a link between asbestos and lung cancers, though again Doll is sceptical of the quality of their research:

I personally don't accept that they established a relationship between asbestos and lung cancer: but they did to their own satisfaction do so, because the German government ruled that ... it was an occupational hazard during the war and the West German government, after the war, continued with the same legislation. But the evidence on which they reached that conclusion would not convince people in this country or the United States.

The evidence was trivial really: it was just seeing a few cases. I always so treasured one sentence in Nordmann's publication in which he described two cases of lung cancer he had seen in asbestos workers, and he said: 'If these had been the only two cases I would have known that asbestos caused lung cancer.' Well, it's utterly stupid! You can't tell from individual cases like that, you have to know what the risk is and how many people were exposed, but ... it was a rather typical way of thinking of classical pathologists: 'I have seen a case under a microscope, therefore there is a causal relationship'; but in England and in the United States you require evidence that the risk in a population is greater with exposure than without the exposure.

Of course there was also evidence to support the claim [of lung cancers being caused by asbestos] based on just seeing a few cases: that it had been possible to produce cancer of the lung in mice by exposing them to asbestos. This was again not an experiment that would have carried much weight in Britain or the United States: it started off with

> about 200 animals of which only ten survived to be old enough to develop a tumour, in which they reported two as having developed a tumour – 20 per cent, as they called it. On microscopic examination by pathologists on this side of the Channel ... one of them wasn't accepted as a cancer, so the evidence again was extraordinarily poor, but it satisfied them. It wouldn't have satisfied people in this country.

Even so, these German scientists were undoubtedly on the right track and had it not been for the intervention of the war and, of course, the revelation of the appalling crimes of the Nazi era, their research might have had a much greater impact than it actually did. Müller's paper did percolate out of Germany during the war – Schairer and Schörniger's did not – and despite its Nazi provenance, who can doubt that it might have spurred public health authorities into some kind of response:

> ...as far as I know, Müller in particular, the first person to publish a paper on lung cancer and smoking, was not connected with the Nazi party in any way... Schairer and Schörniger ... [were] a different matter, they were, of course, in this unit that had been set up by Hitler; but after all, after Müller's paper had been published, it was extraordinary that other people hadn't followed it up, and I'm sure we would have in this country if it hadn't been for the war.
>
> ...there was certainly a backlash to the discouragement of smoking: this was identified very closely with the Nazi regime. Although, I may say, that the actual figures showed that cigarette consumption went up steadily throughout the period. So the public didn't pay much attention to it, and such elderly German scientists that I've known ... that I've asked about that period, said, 'Oh, we didn't pay any attention to it!' But nevertheless it was identified with the Nazi movement, and for years after

the war any suggestion that smoking was harmful was not readily accepted in Germany. And indeed, I think, accounts for the fact that it was a long time before they were willing to discourage smoking to prevent lung cancer.

CHAPTER TWO

life unworthy of life

If the logic of eugenics leads inexorably towards a programme of compulsory sterilization for those with 'hereditary' diseases, then what about those individuals who have 'incurable' diseases or who are so handicapped that they require constant medical attention which will not in fact alleviate or cure their problems? As one might imagine, this was an issue to which the Nazis gave considerable thought.

The issue of euthanasia had been raised in Germany even before the First World War by liberal progressives who argued that welfare spending would be better directed at the healthy workforce rather than at unproductive cripples. This kind of rhetoric was unlikely to evoke much response in a society which still showed respect for Christian values and mores, but this position was to be quickly and systematically changed by the experience of the war.

Germany's participation in and loss of the First World War contributed to a significant change in the moral climate. Michael Burleigh argues: '... it familiarizes people with mass death and it deadens human sensitivities to the sanctity of human life and in this case it also led people to question traditional Judaeo-Christian morality about the sanctity of human life and to adopt much harsher views of the world.'

This was recognized at the time. Karl Bonhoeffer, the chairman of the German Psychiatric Association and father of Dietrich Bonhoeffer, theologian and opponent of Hitler, said in 1920:

> It could almost seem as if we have witnessed a change in the concept of humanity. I simply mean that we were forced by the terrible exigencies of war to ascribe a different value to the life of the individual than was the case before, and that in the years of starvation during the war we had to

> get used to watching our patients die of malnutrition in vast numbers, almost approving of this, in the knowledge that perhaps the healthy could be kept alive through these sacrifices. But in emphasizing the right of the healthy to stay alive, which is an inevitable result of periods of necessity, there is also a danger of going too far: a danger that the self-sacrificing subordination of the strong to the needs of the helpless and ill, which lies at the heart of any true concern for the sick, will give ground to the demand of the healthy to live.

In the same year in which this speech was given, Karl Binding, a lawyer, and Alfred Hoche, a psychiatrist with a morbid interest in brain activity in recently guillotined criminals, wrote a pamphlet entitled 'Permission for the Destruction of Life Unworthy of Life'. Both men were right-wing German nationalists who favoured the idea of loyalty to the 'National Community' above any notion of individual rights. Binding died before the tract was actually published, but Hoche became an aggressive apostle of euthanasia.

The argument advanced by Binding and Hoche followed three main tracks. In the first place they were keen to stress how relatively recent and short-lived the Judaeo-Christian tradition of respect for human life actually was, citing as alternatives the Spartans, who routinely killed weak infants, and the Inuit Eskimos who killed their ageing parents. This led them to suggest that society should kill off 'incurable idiots' as well as the terminally ill and the critically injured.

The second of the three justifications given by Binding and Hoche for euthanasia will be familiar to anyone who has followed more recent debates on the subject: that individuals who are terminally ill or severely injured should be given the right to choose to die on their own terms, through a painless, medically administered procedure. Then, as now, doctors on their own initiative would often ease the final agonies of a terminally ill cancer patient with a massive overdose of morphine, but Binding and Hoche argued that this should be a right, and that doctors who perform this service should not have to fear prosecution as a result. Many people now would agree with this argument, given certain protections, and indeed it has recently been legalized in the Netherlands, but the main thrust of the pamphlet was different.

The crux of Binding and Hoche's arguments was that they believed that certain individuals were 'unworthy of life'. This applied to both people who were so incapacitated by pain or disease that their lives had become unbearable and to those who were so 'inferior' in the eugenic order of rank that that there was no value in their continued existence. In fact the arguments about the terminally ill were little more than a smokescreen: the authors' concern was principally to do away with the feeble-minded and mentally handicapped. Leaving aside their evident personal dislike of the disabled, the arguments they put forward were both eugenic and economic. Binding wrote that: 'If one thinks of a battlefield covered with thousands of dead youth ... and contrasts this with our institutions for "idiots" with their solicitude for their living patients – then one would be deeply shocked by the glaring disjunction between the sacrifice of the most valuable possession of humanity on one side and on the other the greatest care of beings who are not only worthless but even manifest negative value.'

This point was enlarged with enthusiasm by Hoche, who had never recovered from the death of his only son in the battle of Langemarck and who also made his own utilitarian argument:

why should a whole generation of nurses 'vegetate' in the service of these 'ballast existences'? Underlying their arguments was the idea that the 'feeble-minded' or 'idiots' who occupied the asylums were not human. Although they looked like humans and to some extent acted like them, their minds were so profoundly degenerate that they could not really be said to have human personalities: this was a claim which would later be used against Jews.

In conclusion, Binding and Hoche argued that patients, relatives and physicians should be allowed to apply for euthanasia under circumstances in which life was, or had become, unworthy, but that the state should be the final arbiter. Their book caused a considerable stir in Germany (although not in Britain or the USA, where the euthanasia debate remained focused on mercy killing) but received no official acceptance or approval by the government of the Weimar Republic and it was not until the Nazis seized power that it once again returned to the forefront of what passed for the debate on medical ethics under the Third Reich.

In part, euthanasia had become a live issue again because of advances in psychiatry, as Burleigh explains:

> Psychiatry in Germany was not a particularly high-status profession, firstly because it was identified with the abuse of shell-shocked soldiers during the war; secondly because of the misdiagnosis of revolutionaries in 1918–19 as schizophrenics or whatever; and thirdly because economic conditions meant that conditions inside asylums were dreadful.
>
> During the First World War asylum patients were seen as a very low priority, so conditions in asylums essentially deteriorated to the point where ... 70,000 psychiatric patients quietly starved to death...
>
> In the twenties various reformers introduced occupational therapy ... to the point where 80 per cent of patients in asylums performed some

> sort of work, ranging from basket weaving to answering the telephone in the asylum; that's the first reform, secondly they experimented in community care.

Leading these reforms were Gustav Kolb and Hermann Simon, who were quick to point out the economic benefit of what they were doing, as well as the medical improvements. Community psychiatric clinics could successfully treat vastly greater numbers of patients than the grim asylums, at a much lower cost. The effect of these reforms – together with new types of treatment like electro-convulsive therapy and insulin-coma therapy – initially, at least, was to change the mood in psychiatry from pessimism to optimism. The medical profession began to look on asylums and institutes for the mentally ill as hospitals rather than warehouses where the handicapped could be kept out of the way.

But a subsidiary effect of the improvements in psychiatry was to draw attention to the patients who remained resistant to therapy. According to Burleigh: 'The down side of these reforms was that the group of people that were not being cured and that could not work were gradually pushed to the margins of what was already a very marginalized population.'

This brought the eugenic arguments back to the forefront:

> After the First World War people were concerned that during that war there'd been a massive loss of life among fit young males. Germany itself had been subjected to a very debilitating Allied blockade in which people were increasingly starving, and I think there was a general drift to regard human life as cheap.
>
> And then, on top of that, you have ideologues who wished to create a self-conscious new morality, if you like, for a new man and the Nazis are in the

vanguard of that. And part of that new morality for the new man is to go back to the morality of much earlier times. In the Nazi case they wanted to go back to the mores of, say, the ancient Spartans, who allegedly practised infanticide. What they're saying is that our moral weakness, our liberalism, our humanitarianism is compounding rather than solving the problem.

They did certainly claim in absolute statistical terms that more people were in asylums: whether that's true or not ... that is the propaganda claim, you know, there are more and more armies of the unfit: hordes of the unfit.

The essence of eugenics from a psychiatric perspective is that rather than merely presiding over the warehousing of the insane or the ... eugenically defective, that eugenics enables psychiatrists: it gives them a great boost in status and they become the sentinels presiding over the national gene pool so they become immensely important.

The other advantage of eugenics from their point of view is that all those that they can do nothing to help or who in some sense cause them difficulty, they can eliminate so then they are just left with a group of people who they think they can cure, so they've in a way medicalized their function, just as they've converted the asylum into a form of hospital rather than a warehouse.

Even so, fearful of the public row that would be caused, and anxious in any case to prepare the ground first, the Nazis did not actually legalize euthanasia. Hitler certainly favoured the principle. Shortly before the outbreak of war he told Leonardo Conti (the Reich Doctors' leader), Martin Bormann and Hans Lammers (head of the Nazi Party chancellery), that:

He regarded it as right that the worthless lives of seriously ill mental patients should be eradicated. He took as examples the severe mental illnesses in which the patients could only be kept lying on sand or sawdust, because they perpetually dirtied themselves, cases in which

> these patients put their own excrement in their mouths as if it were food, and things similar. Continuing on from that, he said that he thought it right that the worthless lives of such creatures should be ended, and this would result in certain savings in terms of hospitals, doctors and nursing staff.

Before he died Hitler had also told Conti's predecessor as the doctors' leader, Gerhard Wagner, that he intended to implement euthanasia once war had started.

In the late summer of 1939 Hitler ordered the head of the Führer Chancellery, Philipp Bouhler, and Professor Karl Brandt, a surgeon who was a member of his own entourage, to select and authorize certain doctors to begin a programme of 'mercy killing' of psychiatric incurables, together with the registration of all children born with Down's Syndrome, microcephaly and hydrocephaly, missing limbs and cerebral palsy.

In the case of the children, once they had been registered their cases were examined by the Reich Commission for the Registration of Severe Diseases in Childhood. Originally intended to examine the cases of infants up to the age of three, this was eventually extended until the upper age limit was sixteen. Once the selection for euthanasia had been made, the children were transported to a group of specially selected clinics – usually on the pretext of receiving better treatment. There they became a commodity: some were killed by having serious respiratory diseases induced, some with drug overdoses, and some were experimented on before they were murdered. Their parents were usually told that they had succumbed to pneumonia.

One very fortunate survivor of this process has left a harrowing account of her brush with death. Elvira Hempel was born in 1931 into a family of sixteen children, of whom nine died young. Her father was a habitual criminal who made no effort to provide for his children, and rather than attending school they would spend their days picking through the local rubbish tip for scraps of food. After the family became homeless Elvira was left with her grandparents, before being taken into care when her grandfather developed tuberculosis. In 1938 she was declared feeble-minded and taken with her three-year-old sister to the children's asylum at Uchtspringe, where she found herself surrounded by severely mentally and physically handicapped children. She soon realized that the handicapped children were being taken away by someone she called 'der Totenmann' (the Death Man).

On 28 August 1940 the Totenmann came for Elvira and her sister. Elvira was interviewed by a woman who gave her a series of questions to answer and asked her to do a couple of basic puzzles. With this completed – she hadn't done well – she was put in a bus with some of the other children and taken to the prison at Brandenburg to a special euthanasia programme extermination centre. On arriving there she was told to undress so that she could join the children in another room behind a heavy door. This took her a long time as her dress had a lot of buttons to undo. While she was doing this, a man had been studying her file. After a few minutes she was told to dress again and sent to join two other children who would also survive.

For adult incurables, Bouhler and Brandt set in motion a secret service-style operation to ensure that everything ran smoothly. After they had set up a headquarters in a villa at Tiergartenstraße 4 in central Berlin (the operation became known as T-4 after its address), their first problem was the recruitment of personnel. Mostly these came through word of mouth: friends and students of the senior T-4 officials; and doctors of course: this was to be mass murder disguised as a medical procedure. But to the doctors were added police officers and SS men in order to handle the nitty-gritty of killing. One such was Franz Stangl.

Stangl, an Austrian policeman and Nazi Party member, was summoned to Berlin by Heinrich Himmler to receive new orders. Stangl's superior, Kriminalrat Werner, explained what was required:

> ...Werner told me that it had been decided to assign me to a very difficult and demanding job. He said that both Russia and America had for some considerable time had a law which permitted them to carry out euthanasia – 'mercy-killings' – on people who were hopelessly deformed. He said this law was going to be passed in Germany – as everywhere else in the civilized world – in the near future. But that, to protect the sensibilities of the population, they were going to do it very slowly, only after a great deal of psychological preparation. But that in the meantime the difficult task had begun, under the cloak of absolute secrecy. He explained that the only patients affected were those who, after the most careful examination – a series of four tests carried out by at least two physicians – were considered absolutely incurable so that, he assured me, a totally painless death represented a real release from what, more often than not, was an intolerable life.

Stangl eventually accepted and became 'security officer' at Hartheim castle, one of the main killing centres of the euthanasia programme.

To understand the full horror of this programme, it is worth examining the way in which murder became a normal and accepted part of ordinary medicine under the Third Reich. In the

1980s Dr Michael von Cranach started work at the Kaufbeuren Hospital in Bavaria and did just that:

> When, in the late 1970s, reform of psychiatric services started in Germany, many of us – then young psychiatrists – went to big mental hospitals in order to change the whole psychiatric system. When I started here, in 1980, I soon realized that the terrible past of this hospital was, in some way, present continuously. I knew, before, that things like that had happened, but I had no concrete idea of what it was. And here I found, when I started asking people what happened here, that they soon told me stories. So I found out that the past was not buried – or really a past – but it was present and so we started to go into the archives and look what happened with the patients.
>
> And as soon as we started doing this openly, very many people working here came to us, and to me, telling me, 'Well, my father has worked here and I know this and that story.' And relatives from patients started coming and asking, for the first time, what happened to their patients who had been killed thirty years or forty years before.
>
> So, I and all the other people who were in similar situations, realized that the psychiatric reform could not be accomplished without really looking into the past and making it open and discussing it.
>
> The first thing I discovered is that I went to the admission books, and discharge books, and I found that so many hundreds and thousands of patients died here, in that time. So, we went into the archives and looked at the concrete case notes. We started reading the little literature which was available at that time. And so we slowly started to know in more detail what happened.
>
> We found out that this hospital was one of the ... more terrible hospitals in the way that here, not only in the first part of the euthanasia programme patients were sent to the special killing hospitals. But that

> after '43, when this programme was stopped, patients continued to be killed here, either by the terrible diet through which the patients died of [malnutrition]. And, later the patients were killed by injections given by nurses and doctors: all organized and planned. And then we found out that medical experiments were performed on patients, and then we learned later that forced labourers were sent to psychiatric hospitals, even to this one, in order to either get cured immediately, and if there was no solution to their problem they were killed. And so the more we looked – went into the archives and investigated – the more we found out about this past.

One of the surprising aspects that von Cranach discovered was how widely the local population had been aware of what was taking place at the hospital:

> Oh yes, naturally they knew ... but it was a thing which nobody dared to speak about. And when we started publishing about these events, we got enormous feedback from people telling us personal stories of what they heard, and the priests of the little villages came and told me all what they knew and what they experienced at that time. People came, saying how in their small way they tried to counter-attack all these things. So, we found out that ... everybody was aware of what happened here, but nobody dared to speak about it.
>
> ... Investigating in our archives, we found a whole, big file of something like 250 letters, which were collected by my predecessor; letters of relatives ... either asking what happened to the patient ... the relatives of these patients were notified by just a very simple letter, saying that your relative has died by pneumonia or whatever, please don't ask about the fate of the corpse, [it has been] cremated and so on. And in responding to these letters, many relatives asked in the letter, 'Please tell me the truth. I know about this problem... I put my relative into your medical

> responsibility and why did you kill him?' And so on. So ... most of them are very, very critical, but many are in favour: 'We thank you for having saved us from this burden.'

It is quite likely that this was the result of years of eugenic propaganda: '...this was the aim of the propaganda, but I was astonished to see that most of the letters were critical.'

For all that the process was supposedly intellectually justified, it was nevertheless hedged around with lies and evasions:

> The first ... patients were sent to special institutions. There were six institutions in Germany, where they had installed these gas chambers, which later were transferred to the concentration camps. And the patients [were collected] ... with a bus service from the hospitals to bring them to the special clinics. And ... some hospitals, which were, perhaps, more open to the programme, this one, patients from other institutions were sent, before they were sent to these final extermination hospitals, in order to blur a little bit for the relatives the way the patient took. So, for instance ... many patients from Hamburg, from Berlin, from everywhere, were sent here, and from here to another hospital and then, finally, for us it was Gräfeneck, which is nearby here, or to Linz, which is now in Austria.
>
> We know, in total, that in this first episode ... something like 70,000 patients were killed. From here, about 900 patients were sent. And all the places which were empty, because the patients were sent away, were filled up immediately with patients from other hospitals which were later, again, sent to these extermination hospitals. And there, the doctors – there was, in these extermination hospitals, always a doctor, seeing the patients before they entered this gas chamber. And this doctor wrote a death certificate, which was sent to the relatives – 'we notify you that your relative has died of pneumonia or appendicitis', and so on. But the relatives, naturally, knew.

There are some letters saying, 'My son had his appendicitis operations twenty years before, how could you tell me such a lie?' And things like that.

Although, in theory, the extermination was supposed to be a painless and untraumatic 'medical' procedure, the reality was much different:

The hospital was informed, with very short notice, mostly the day before, that the next day would come a bus and collect eighty persons, or so; and with a list of these patients. We know, from nurses who witnessed this later, all these events, that some nurses tried either to hide persons, or ... they contacted the relatives and asked them to take the patients home, which happened in very few cases.

But the lists came, and then the patients were prepared; this was all by advice of the central administration in Berlin. They had to get their sticking plaster on their skin, on the back, with their name and their number. And so they were put into the buses. We knew here, concretely, the story that when the first bus came away, patients were quite normal and behaved quite normally, because they didn't know what was happening to them. But a week later, all the clothes of the patients were sent back from Gräfeneck to Kaufbeuren, which were full of vomited things, in very terrible conditions, so that the nurses and everyone knew, from this, that the patients had been killed.

Naturally, there was gossip about it. But now they knew for sure what happened. And, naturally, the nurses told the patients, so the patients knew. And so when the second transport came, the patients were very, very ... desperate, in a terrible situation, and several cried and – and there is a witness which said that one patient wanted to confess and to speak to the priest. And some patients had to be attached with chains to their seats in the bus because they did not want to come with them.

So they were brought to Gräfeneck and immediately they were received by a doctor and a couple of nurses, and these had the case notes. Then they verified: 'this is the patient with this patch on the back, [who] was the patient who was described in the case, who was on the list', you know. And then he was undressed and then he was told that he had to join a room where there were showers, and when something like twenty or thirty patients were together, then they closed the doors and they just introduced carbon-monoxide gas into the room, and the doctors through a window observed the dying of the patients.

And when they all were dead, they opened the room and ... in some of the places they were buried and some they were cremated. And then the relatives got this mock information about the death of the patient.

Burleigh quotes the final letter of an 'incurable' epileptic woman to her father, a retired doctor, in October 1940:

Dearest beloved Father!

Unfortunately it cannot be otherwise. Today I must write these words of farewell as I leave this earthly life for an eternal home. This will cause you and yours much, much heartache. But think that I must die as a martyr, and that this is not happening without the will of my heavenly redeemer, for whom I have longed for many years...

I embrace you with undying love and with the firm promise I made when we last said our goodbyes, that I will persevere with fortitude.

Your child Helene

On 2 October 1940. Please pray a lot for the peace of my soul. See you again, good Father, in heaven.

So much for hopeless 'idiots'.

Although Nazi propaganda always characterized 'incurables'

as ghastly, frightening mutants at best, vegetables at worst, the reality was that most of them were not, and were much-loved family members. Under these circumstances, it was inevitable that resistance to, and protest against, the euthanasia programme would grow. Hans-Walter Schmuhl explains what happened:

> There was a surprisingly large amount of resistance. There were repeatedly demonstrations outside the killing centres, something completely unusual for the Third Reich, also something that did not happen at all with regard to the persecution and extermination of the Jews.
>
> That really does require an explanation, because we have to assume that the prejudices against mentally ill and mentally handicapped people, which exist everywhere and at all times, also existed in the Third Reich, and that they were more likely to have been strengthened by the National Socialists' continuous propaganda.
>
> If we look more closely at what people have said, we can see that there was not so much a fundamental criticism of the killing of disabled and mentally ill people, but that the criticism and the opposition were directed very strongly against the way in which the killing was carried out: that there was no law, that the relatives were not informed, that it was a process completely without rules, unfathomable, and so on.
>
> That is, of course, connected with the fact that potentially everybody can come to need psychiatric treatment and be threatened by extermination as a result of developing a geriatric illness, say, or having an accident, or as a soldier at the front suffering a brain injury, and so on.
>
> And so one could not, as in the case of the persecution of the Jews, stand to one side and say: 'Well, I'll never be persecuted!'
>
> But instead, everybody was affected, and so there was this great uneasiness. The National Socialist regime, or rather the top ones in the regime, also took account of this.

However, the most prominent opponent was the Roman Catholic Bishop of Münster, Clemens August Graf von Galen. He was a member of an old aristocratic family: ultra-conservative, snobbish, racist and reactionary; but when he learned of T-4 and the euthanasia programme, he denounced it ferociously:

> We are not dealing with machines, horses and cows whose only function is to serve mankind, to produce goods for man. One may smash them, one may slaughter them as soon as they no longer fulfil this function. No, we are dealing with human beings, our fellow human beings, our brothers and sisters. With poor people, sick people, if you like, unproductive people. But have they forfeited the right to life? Have you, have I the right to life only so long as we are productive, so long as we are recognized by others as productive?

Galen's sermons attracted attention outside Germany as well as inside. They were reported by the BBC and in British newspapers, and copies were dropped by the RAF as propaganda pamphlets, but such was his status and popularity among the devout that the Nazis were wary of moving against him while the war continued. Even so, Hitler had registered a personal grudge against him, and was still apparently brooding upon it a year later:

> The fact that I remain silent in public over Church affairs is not in the least misunderstood by the sly foxes of the Catholic Church, and I am quite sure that a man like the Bishop von Galen knows full well that after the war I shall extract retribution to the last farthing. And, if he does not succeed in getting himself transferred to the Collegium Germanicum in Rome, he may rest assured that in the balancing of our accounts, no 'T' will remain uncrossed, no 'I' undotted!

Even so, Galen was not the only religious leader to protest. The Bishop of Limburg also made his views known in a petition to the government:

THE BISHOP OF LIMBURG LIMBURG/ LAHN, AUG 13, 1941

To the Reich Minister of Justice *Berlin*

Regarding the report submitted on July 16 (Sub IV, pp 6-7) by the Chairman of the Fulda Bishops' Conference, Cardinal Dr Bertram, I consider it my duty to present the following as a concrete illustration of destruction of so-called 'useless life.'

About 8 kilometres from Limburg, in the little town of Hadamar, on a hill overlooking the town, there is an institution which had formerly served various purposes and of late had been used as a nursing home; this institution was renovated and furnished as a place in which, by consensus of opinion, the above mentioned euthanasia has been systematically practised for months – approximately since February 1941. The fact has become known beyond the administrative district of Wiesbaden, because death certificates from a Registry at Hadamar-Mönchberg are sent to the home communities. (Mönchberg is the name of this institution because it was a Franciscan monastery prior to its secularization in 1803.)

Several times a week buses arrive in Hadamar with a considerable number of such victims. School children of the vicinity know this vehicle and say: 'There comes the murder-box again.' After the arrival of the vehicle, the citizens of Hadamar watch the smoke rise out of the chimney and are tortured with the ever-present thought, of the miserable victims, especially when repulsive odours annoy them, depending on the direction of the wind. The effect of the principles at work here are: Children call each other names and say, 'You're crazy; you'll be sent to the baking oven in Hadamar.' Those who do not want to marry, or find no opportunity, say, 'Marry, never! Bring children into the world so they can be put into the bottling machine!' You hear old folks say, 'Don't send me to a state hospital! After the feeble-minded have been finished off, the next useless eaters whose turn will come are the old people.'

All God-fearing men consider this destruction of helpless beings as crass injustice. And if anybody says that Germany cannot win the war, if there is yet a just God, these expressions are not the result of a lack of love of fatherland but of a deep concern for our people. The population cannot grasp that systematic actions are carried out which in accordance with Par. 211 of the German criminal code are punishable with death! High authority as a moral concept has suffered a severe shock as a result of these happenings. The official notice that '*X.*' had died of a contagious disease and that for that reason his body has to be burned, no longer finds credence, and such official notices which are no longer believed have further undermined the ethical value of the concept of authority.

Officials of the Secret State Police, it is said, are trying to suppress discussion of the Hadamar occurrences by means of severe threats. In the interest of public peace, this may be well intended. But the knowledge and the conviction and the indignation of the population cannot be changed by it; the conviction will be increased with the bitter realization that discussion is prohibited with threats but that the actions themselves are not prosecuted under penal law.

Facta loquuntur.

I beg you most humbly, Herr Reich Minister, in the sense of the report of the Episcopate of July 16 of this year, to prevent further transgressions of the Fifth Commandment of God.

Dr Hilfrich

In the event the euthanasia programme was suspended in August 1941. The original plan had been to eliminate some 70,000 patients and thus to free up resources in the hospitals to look after curable cases and military casualties, as von Cranach explains:

> Yes, in 1941 this first phase of the euthanasia programme was officially finished. There is much historical [debate] about why this was the case. There was, there's no doubt, quite a lot of open protest against it, from especially the Catholic Church. There was a famous speech of Cardinal Galen in Münster, and there was even protests from within the party. There is a letter of Himmler saying that one should stop it, because the population is nervous and it's not really accepted by everyone and so on. But other historians say that these were only side factors, that the main factor was that they thought, at that time, that they had accomplished the task of this programme.

Instead a new system was implemented:

> But then a second phase started, which the historians call the 'wild euthanasia'. So the killing continued in the hospitals itself. And not only in Kaufbeuren: in most of the other psychiatric hospitals, with different degrees of intensity. What happened? Here in Bavaria we know that there was a meeting in 1941 where the Ministry of Health in Munich summoned all the psychiatrists, directors of the hospitals, and the director of Kaufbeuren and proposed the introduction of ... a completely fat-free diet with more or less no calories, which was calculated in such a way that the patients, after four or five months, would die of starvation, or the previous consequences of starvation. And then they decided to introduce this diet in all psychiatric hospitals in Bavaria and in other parts; then it was introduced in other parts of the Reich, too.
>
> ... not all the hospitals introduced it with the same intensity. We know

that some hospitals tried to avoid introducing the diet. And we know, now, that it was after the war, many said, well ... this was consequences of the war, because of the lack of food for everybody; but that's not true, there was enough food. And we know from these conferences that their aim was just really to continue to kill those selected patients which they thought were not worth living. So this was one way of continuing the problem.

The second was that in many hospitals they started, with the help of the central administration in Berlin [i.e. T-4], killing patients themselves. Here [at Kaufbeuren], the director asked the central administration to send them nurses experienced in the killing of patients. And two nurses came here, and they opened two wards: one ward for men and one for women; and then they started selecting patients from the different wards, transferring them to that ward, and here they were killed with injections of opiates and scopolamine. And we know that about six hundred patients were killed personally by these special nurses and the doctors who supervised them.

Another form of killing was that in several hospitals ... they installed special divisions for children. And to these wards, children – mostly mentally retarded children from charitable organizations – were sent to the hospital and killed. And here [at Kaufbeuren] we think that something like 250 children have been killed.

Nowadays we have become to a large extent blasé about the extent of the crimes of the Nazi era, but one of the more macabre and fascinating aspects of T-4 and the euthanasia programme is the way in which it was conducted as a form of medicine, and, despite the presence of SS personnel and policemen, administered by doctors with no obvious reason to be enthusiasts for mass murder. Von Cranach says:

The most uncomfortable thing is these were not in some way ... sort of psychopaths or people who, in the Third Reich, had come to power who were not representative of the normal psychiatrists. They were all good, humanistically trained doctors. And we have to try to find out what happened that they changed to this direction. We know that the director here, Dr [Valentin] Faltlhauser, he was a very prominent psychiatrist of his time. In the twenties, we had a very positive development of psychiatry in Germany. There was the famous Dr Simon in Gütersloh who developed work therapy and occupational therapy... And in Erlangen there was a world-famous Dr Kahl, who introduced, for the first time, principles of community psychiatry in the way that he discharged patients from hospitals and tried to organize community help, in different forms: work, rehabilitation, shelter, accommodation, psychiatric therapy, even psychotherapy and things like that. And this was very, very, progressive. Patients – people – professionals from all over the world came to visit this thing. So, Dr Faltlhauser was the main assistant or doctor who organized this, so he had written a lot of articles about this new form of community-oriented psychiatry.

But then you can see that, at the end of the twenties, at the beginning of the thirties, before Hitler came to power, you can see how this community psychiatric movement more and more changed from the basic philosophy. At the beginning it was really helping individual persons, but more and more eugenic concepts came into the philosophy, and the way that they were more interested [in using] this community psychiatry to control society, and to control the health of the population.

The emphasis changed from, 'I am a doctor and responsible for individual patients,' to, 'I am a doctor and I am responsible for the nation's health', which is a very dangerous step, and one of the main criticisms one has of the eugenics movement.

One of the ways in which eugenics neatly dovetailed with the philosophy of Nazism is clear:

And so at this time Dr Faltlhauser became director of Kaufbeuren [which coincided with the increased discussion of euthanasia in professional journals]... And Dr Faltlhauser was much against this movement and he wrote then an article, very emphatically that this is the wrong way to deal with the problem and criticized these thoughts very, very intensively.

But in 1933 soon after the taking of power of Hitler, the forced sterilization laws were introduced. Here, you see the first change in Dr Faltlhauser's attitudes. He became ... the regional expert, to appear in court and to initiate the sterilization process for the patients, and he got very much engaged in this. Why? That's difficult to say. It's really very multi-factorial. First thing, I think it's because this gave importance to his person and to his job. Secondly, maybe even one has to think that this was connected, naturally, with money: you got money for all this expertise.

So now you can see that eugenics, or eugenic thinking, appears here, in all the case notes, for the first time. Words like 'the health of the Volk' – of the people – 'is the main goal of our action', and things like that.

And soon, like in all other parts of Germany, very many patients were admitted to hospital because if those who should be – according to the law – sterilized, refused to be sterilized, they were sent to the psychiatric hospital, even not having any acute reason for being treated, and they were kept in the hospital until they, more or less ... agreed to be sterilized. And so many, many patients were admitted to the hospitals.

And then, the next thing which happened is that in the middle of the thirties they reduced enormously the funds they diverted to the hospitals. The economic situation of the hospitals became very, very poor. With the explicit formulation that there is no need to spend so much money on these patients ... for the first time now, you see an argument coming out that the psychiatrists protested against this reduction in funding. And they said that psychiatry needs money to treat the curable – and by reducing the funds for the hospitals, they hinder the hospitals' treatment of the curable.

They understand the [necessity for] reducing funds, but this should only be directed, in some form, at the incurable. And here, for the first time, comes this differentiation that ... psychiatry has patients which they can cure and should invest enormous efforts in curing them; and that there are patients which are incurable, without saying what should be done with these.

And the answer to this question came, in 1939, with the official euthanasia programme.

It is undeniably difficult to fathom the psychological gymnastics which the euthanasia doctors must have undertaken to turn from attempting to care for patients towards simply liquidating the difficult cases, but von Cranach is contemptuous of at least one recent explanation:

Well, it is very difficult for us to understand all the factors which brought these people to act in this way. There is one factor which is very much discussed in Germany, [although] I have my doubts that it is so important. It has something to do with the problem of curable and incurable; and that is that the doctors felt that the act of killing was a therapeutic act: that they help the curable to get cured, and they help the incurable – mercifully – to overcome all this terrible fate, to be a psychiatric chronic patient. So killing as a therapeutic act.

I don't believe so much in this interpretation, because if you read the [testimony] and hear the witnesses, how they killed their patients, and how the situation of the hospital was, you can't really see, in any way, signs of pity or mercy... It was really a terrible form of killing patients, to let patients die from starvation and observe them over months. That cannot be something to do with: 'I want to help this person to overcome his pains and his terrible fate.'

In reality, he argues, the chief factors at play were more to do with the psychology of the doctors than practical considerations about their patients:

You must realize that all these terrible things only happened in twelve years, from '33 to '45, so there was an atmosphere of, 'We are really in revolutionary times and now we are free to do whatever we think is good...' And we are members of this group which is doing the good thing for humanity.' So this was one factor.

The second factor is something which we call 'anticipating obedience'. So that they know what the goals of the system were, but they were free to interpret them and to act, according to them, and even to go further, to do it better than expected from them. So they made – continuously – proposals how you could solve this problem more efficiently, and so on, in order to be more involved in the inner circle and get the admiration of the system.

And another aspect – which has something to do with it – is that the whole organization was psychologically very, very efficient. For instance, this euthanasia administration in Berlin had a sort of guest house or hotel on a very beautiful lake near Salzburg, and all the directors were continuously invited to it with their wives, and they had champagne and wine and good food – food, which, at that time, was not accessible to most of the people. They danced. And we know this from the very detailed diary of one of the doctors, Dr Mennecher. And this made them all members of ... an elite.

At the same time, they realized that this was wrong. We know from Dr Faltlhauser that he went here, to the local Catholic priest, and told him,

> 'You know that in the hospital terrible things are happening, and if you are right and we will all appear ... in front of our creator, we will be punished for having done that.'

It was also certainly the case that, as the action continued, the victims of euthanasia became increasingly dehumanized as far as their 'carers' were concerned. This was parallel to what happened in the concentration and extermination camps, although perhaps more surprising, as concentration camp staffs were largely made up of former policemen and low-grade Waffen-SS personnel, with little or no experience of caring for others. But for doctors and nurses, who had hitherto put all their efforts into looking after their patients, it must have required a considerable effort to change their outlook:

> This, naturally, is right, that when you see people living under such poor conditions, and so dehumanized and degraded that it's easier to make the decision to kill them. But, again [we had] a sad experience, going through the notes of the patients ... we can see that, until 19... perhaps '39 or '40 or so, patients are described in the case notes ... with the psychiatric vocabulary used all over the world. And then, the descriptions change.
>
> And they do not change into descriptions of misery [and] you can't see any signs of pity; or seeing they are so poor or they're so miserable; or there's no chance of healing them or so on. What you see is an attribution of value – they start valuing their patients, you know. For instance, saying they are lazy ... they are unclean; giving the patients the responsibility for their misbehaviour, in the view of the psychiatrist.

By transferring the blame for their plight to the patients, despite the fact that their illnesses and disabilities were supposedly hereditary, the doctors and nursing staff could prepare themselves to perform the intellectual somersault which would allow them to murder them:

> There's a [case] which we looked into, in very much detail. This is the story of Ernst Losser, [who] was a boy of ... 10 or 11 when he was referred to Kaufbeuren. And the story of this boy reflects the whole terrible situation at that time. He was the son of a person who did not work regularly and who was selling things from house to house. And in the village where he lived ... some people said that he may be of gypsy descent. So the father was looked at as a difficult person, and he was transferred to Dachau. And the children – there were three siblings – were transferred to orphanages.
>
> And in this orphanage, this boy behaved very [badly]. He tried to avoid visiting the school, and he stole things from others, and so ... he was difficult to educate.
>
> And so they asked for a psychiatric [evaluation]; a psychiatric expert from here came, made an [evaluation] said that this boy is unchangeable and no educational means will change his behaviour, so he has to be transferred to the children's unit at Kaufbeuren. And here, this boy was not psychiatrically ill, nothing: you know, he was not even mentally retarded; he was a difficult boy in a difficult life situation.
>
> And we know, from the witnesses, from nurses afterwards, speaking about him, saying that he was a very, very friendly and completely normal boy, and they liked to play with him and ... one said that he had taken him several times home for a day, because they all had pity with him and thought that he was on the wrong place, even in their view.
>
> There is a very interesting story, or a moving story ... several witnesses have told this story – because he was living with patients who were under this starvation diet, he went several times to the kitchen and stole things: apples and bread and so on; and distributed them to the people who were

on the starvation diet. And this came to the knowledge of the psychiatrists and of the administrator, and so they decided that ... this boy was a nuisance and had to be eliminated. And there were several nurses who said, 'No, this boy is different, we don't think that he should be eliminated.' But nevertheless, they killed him with injections. And, as they noticed that there was some sort of dissent in the hospital about him, the administrator participated in the killing.

So this is the one side of the story. If you now read the case notes, then you see that this story is completely confounded, told in a very different way. It says that this boy is a liar, that he is a thief, that he continuously goes to the kitchen and takes things away and so on. So, he's described as – they say it openly – he is a 'bad character.' So they have abandoned medical vocabulary and ... a medical problem, and they make value judgements on him as a person. This makes it easy: it's easier to kill a thief than to kill a patient.

Another insidious pressure was the fact that euthanasia opened up new areas for research:

The euthanasia action opened up completely new possibilities that had never existed before: now it was possible to couple pathological examinations and clinical observations with each other, and many pathology institutes fell victim to the temptation to exploit this possibility. I myself have concerned myself with the Kaiser Wilhelm Institute for Brain Research, and have attempted to write a history of the institute, making clear how there was a drift in brain research into the area of murdering ill people.

It can be seen that a change in the Head of the Institute in 1937, when Oskar Vogt was forced to resign his office and Hugo Spatz came in his place, also led to a change in the concept of brain research, away from the healthy brain to the sick brain. Questions about the pathology of mental illness and disablement predominated, and the practical consequence of this was that there was a network within institutional psychiatry in Berlin, Brandenburg, and also on a national level. Personal networks came into being here, which then, as the euthanasia action continued, led to close contacts with the extermination process. The most important of these contacts was made by Julius Hallervorden.

He was the Head of Pathology of the Brandenburg mental homes, which finally were located in Brandenburg-Görden, and in 1938 this pathology department became part of the Kaiser Wilhelm Institute for Brain Research, and Hallervorden became Head of Department and Deputy Director of the Institute.

Because Brandenburg-Görden became a centre of the euthanasia action, one of the two research and observation departments was set up here, it became virtually automatic for brains from euthanasia victims to come via the pathology section to the Kaiser Wilhelm Institute for Brain Research. But that was certainly not the only contact that this institute had with euthanasia; brains were sent to Berlin by many doctors who had sat in on classes since 1937 and who had scientific contacts with the Institute, and who from 1940 were themselves involved in murdering patients in the death institutes or were involved in this action in some way.

It wasn't at all necessary for Julius Hallervorden and Hugo Spatz to ask for brains; there were many colleagues who knew about their research interests and in interesting cases made sure that, after the patients were murdered, the brains were sent to Berlin, so that the research could be continued.

Julius Hallervorden went to the Brandenburg killing centre, that is the former prison in Brandenburg, and removed brains from children who had just been killed and took them to Berlin. This is, so to speak, the most radical form of the cooperation that one finds here.

One asks, of course, how anyone who began by practising psychiatry could be capable of such a deed. I asked myself that question and found in the documents that Julius Hallervorden, like other brain researchers too, who, when they thought about the ethical implications of what they were doing, divided the whole process into small steps. So they therefore didn't concern themselves with how the patients died, because that didn't fall within the area of their responsibility; instead, they concerned themselves with their professional ethical considerations.

They said, these children are dead, and either they can be cremated immediately, or I can take the brains and examine them and make scientific progress, and perhaps in this way contribute to diseases and disablements in future being able to be alleviated or cured.

Of course, extending the logic of this, one might easily persuade oneself that, because they conducted experiments on the brains of murdered children, these patients had not died entirely in vain:

And faced with these alternatives, they decide to cooperate. But that is only possible if they themselves reject all responsibility for the entire process, and point to the general responsibilities ...

There's another aspect: after 1939 the Kaiser Wilhelm Institute is very strongly militarized. Three special organizations are set up, which are occupied with brain damage and Air Force problems, and as a result the scientists, actually against their will, are directed to deal with other problems.

The euthanasia research provides them with the opportunity to work on subjects and questions that were of interest to them before the war, in particular that applies to Julius Hallervorden, questions about distinguishing between innate and inherited feeble-mindedness. Incidentally, it is also the other way round, that the areas of interest that were pursued in the Brandenburg research and observation station were partly influenced by the work at the Kaiser Wilhelm Institute.

I have discovered from the documents that in [the hospital at] Brandenburg-Gördern they were also interested in so-called athetoses,

> that is, neurological disorders associated with involuntary movements. That is something that they brought as an interest from Berlin; it wasn't originally the interest of the researchers who were involved in the euthanasia action. And they thought about the extent to which such patients should be included in the euthanasia action. So, one can see, there is a research connection, and there is a mutual influence.

One of the reasons that the T-4 project was wound up was that the expertise of its administrators was needed elsewhere. In early 1941 Himmler had approached T-4 to arrange to use the gassing facilities to liquidate incurably sick inmates at the concentration camps, and while this did indeed include many of the camps' Jewish inmates, this was before the systematic extermination campaign began. But this was certainly in the back of the SS leader's mind, and had been for some time.

In fact the SS had begun to implement the Nazis' policy of extermination of their opponents – real and imagined – on the heels of the German invasion of Poland in September 1939. Five Einsatzgruppen (Task Forces) had followed the military into the country, ostensibly to set up a security and intelligence infrastructure for the occupation, but, at least in part, with the express intention of eliminating the Polish 'ruling classes'. Armed with pre-established lists, the SS men had rounded up teachers, doctors, government officials, priests, landowners, aristocrats and businessmen, most of whom were executed by shooting in hastily organized concentration camps and detention centres on Polish territory. The Einsatzgruppen also began the task of identifying and herding Jews into ghettos and camps, prior to beginning their exploitation as slave labour. Two years later, as the German army stormed into central Russia, heading for Moscow and Leningrad, again with Einsatzgruppen in its wake, an altogether more murderous policy was in the early stages of implementation.

The precise date on which the orders for the extermination of European Jewry were given remains obscure: it was probably during the summer of 1941, around the point at which the Blitzkrieg was unleashed upon Soviet Russia. Certainly a warning order was issued by Hitler in his general directive to the chief of the Armed Forces High Command (OKW) in March 1941, when he assigned the responsibility for dealing with the 'Bolshevist/Jewish intelligentsia' to Himmler. This was followed by detailed planning by the chief of the Reichssicherheitshauptamt (RSHA, the Reich Security Main Office), Reinhard Heydrich, and the organization and training of new Einsatzgruppen to follow the army into Russia and begin the murder.

As had been the case in Poland, the first wave of killings was conducted with simple, straightforward brutality. An Einsatzkommando, a sub-unit of an Einsatzgruppe, would round up as many Jews, communists or other target group as they could from whichever locality they happened to be in, take them off somewhere secluded, often make them dig their own graves and then shoot them. Although effective, this was not a popular system with the murderers themselves, except in the relatively few cases where they were actually sadists or psychopaths. As Paul Blöbel, one of the Einsatzgruppen commanders explained, with no apparent sense of irony: 'The nervous strain was far heavier in the case of our men who carried out the executions than in their victims. From the psychological point of view they had a terrible time.' It was also expensive in terms of ammunition, and labour-intensive. The intention now was to eliminate at least a million Polish Jews, many times more Russian Jews, together with the usual roster of other unfortunates: gypsies, communists and so on. Himmler summoned Police Inspector Christian Wirth, of T-4, to resolve the problem.

With his experience of supervising the elimination of more than 70,000 incurables, Wirth was clearly the man for the job. He assembled a group of his experts from T-4 and used them as the nucleus for the foundation of a series of camps to be built along the River Bug in occupied Poland. There they set up gas chambers, similar to those they had used in Germany, where they were able to pump carbon-monoxide fumes, diverted from diesel engines. In this way, at the Belzec camp situated on the Lublin to Lwow railway, for example, Wirth was eventually able to supervise the murder of 15,000 people every day; at Treblinka, Wirth's most effective camp, which was briefly under the command of Irmfried Eberl, one of the T-4 doctors, the capacity was 25,000 people each day.

Nevertheless, despite these almost unbelievable figures, a dispute arose between the T-4 specialists and the technocrats of the SS over the most effective way of killing their victims. At the slave labour camp at Auschwitz during 1942, the commandant had been experimenting with a new killing agent, the prussic acid-based Zyklon B, originally a by-product of Fritz Haber's work in the First World War and now being marketed by the Degesch company as a pest-control fumigant. (Zyklon was made with a special indicator smell added in order to warn users of its presence and thus prevent unfortunate accidents, whereas Zyklon B lacked this.) Zyklon B was more convenient to use – a fatal dose for a large group of people came in the form of a small tin of pellets – and was much quicker than carbon-monoxide poisoning. Consequently trials were arranged to compare the two systems. One of these is described by SS-Obersturmführer Kurt Gerstein:

> The train arrives; 200 Ukrainians fling open the doors and hunt the people out of the trucks with ox-hide whips. Instructions come through the loudspeaker: strip completely, including artificial limbs, spectacles, etc. Then the women and girls go off to the barber, who, with two or three strokes of his shears, cuts off all their hair and stuffs it into a potato sack. Then the procession starts, an extremely pretty girl in the lead. So they move down the alleyway, all naked, men, women and children, artificial limbs gone... They mount the steps, hesitate and enter the death chambers. The majority say not a word. One Jewess, aged about forty, eyes blazing, calls down upon the murderers the blood which is being shed here. She gets five or six strokes of the whip across the face and disappears into the chamber like the rest. The chambers fill up, packed tight – that's Wirth's order. People are treading on each other's toes ... at last I understand why the whole set-up is known as the Heckenholt system. Heckenholt is a little technician who drives the diesel motor and also constructed the installations. The victims are to be killed by the diesel's exhaust fumes.

However, on the day of the trial the Heckenholt system went awry: the diesel engine refused to start, as Gerstein reports:

> Wirth arrives. He is clearly extremely embarrassed that this should happen just on the day when I'm here. Yes, indeed! I can see the whole thing! And I wait. My stopwatch ticks faithfully on – 50 minutes 7 seconds and still the diesel won't start! The men in the gas chambers

> are waiting. In vain! One can hear them sobbing and weeping. 'Like in the synagogue,' says SS-Sturmbannführer Professor Pfannenstiel, Professor of Health Studies at the University of Marburg, as he puts his ear to the wooden door to hear better.
>
> Wirth strikes Heckenholt's Ukrainian assistant twelve or thirteen times across the face with his riding crop. After 2 hours and 49 minutes – measured exactly on the stopwatch – the diesel starts ... a further 25 minutes pass. Correct – many are dead. One can look through the peephole when the electric light in the gas chambers is on for a moment. A few are still alive after 28 minutes. Finally after 32 minutes they're all dead. Men of the labour detachment open the wooden doors from the other side. Jammed in the chambers, the dead are still standing there like marble pillars. There's no room for them to fall or even bend over.

This was, of course, the same 'therapeutic' procedure used in the euthanasia programme.

Finally, it is instructive to record how committed the doctors were to the euthanasia programme. There is no better illustration than the situation described by von Cranach:

> The wild euthanasia went on in all hospitals, up to the point when the Americans or the English or the French came and occupied the hospitals. And in Kaufbeuren [because] ... the doctors told the Americans that there was typhoid fever in the hospital, the Americans did not dare to enter. So patients continued to die and to be killed. And it was only when the Americans slowly learned what happened that after three or four weeks they came to the hospital and they found that, a few days before, the last patient had been killed. Richard Jenner, a four-year-old boy ... it was carried out three weeks after the end of the war.

Why this had happened is difficult to imagine: 'Well, I can't give you a good answer. I can imagine that they had lost any sort of critical control over themselves and that they went on because they thought that this was a solution to many problems; and I can't tell you what happened inside them. It may have much more simpler explanations, that – perhaps – they wanted to get rid of witnesses; this may be the case, but we don't really know...'

CHAPTER THREE

further, higher, faster

'What the *fuck* was that?' On 13 May 1944, 25,000 feet above northern Germany, something unexpected had just happened: a small, red aircraft, without a tail, had streaked through a formation of American B-24 Liberator bombers at an amazing speed, cannons thumping. Gunners in the US aircraft had attempted to track the plane without success: its speed was well beyond anything they had been trained or experienced to deal with; escorting P51 Mustang fighters were left standing by a German aircraft which could travel 200mph faster than their maximum speed. It seemed, for a time at least, that the balance in the air war over Europe might be about to alter. The German Luftwaffe had just used the first operational rocket-powered aircraft, a machine far in advance of anything available to the Allies: the Messerschmitt 163b Komet.

In fact at the outbreak of the Second World War the German Air Force, the Luftwaffe, was probably the most feared weapon in the German armoury. The destruction of Guernica in the Spanish Civil War, followed by the almost unbelievable success of the German armed forces in Poland and in the Western Campaign of 1940 appeared to confirm this, but then came the Battle of Britain. Flying against the numerically weaker but better organized Royal Air Force, the Luftwaffe saw its many weaknesses exposed. It had been set up primarily as a tactical air arm designed to operate in support of ground forces. This it did superbly well: a good proportion of the success of German blitzkrieg tactics was due to the fact that they were carried out in three dimensions. But the Luftwaffe was ill-equipped to perform at the strategic level: the bombers lacked the ability to carry heavy loads long distances, and the fighters, while they matched Britain's Spitfire and Hurricane fighters in terms of performance and firepower, lacked the range to effectively escort the bomber force.

More importantly, however, the strategic direction of the Luftwaffe was profoundly flawed: bombing targets in the Battle of Britain varied between coastal convoys, RAF Fighter Command's advanced airfields, the RAF sector stations, Bomber Command airfields and industrial targets, while no particular effort was made to destroy, for example, the early-warning radar system.

By contrast, a strategic air offensive had figured in British plans for future wars from the time of the First World War. German attacks on London in 1917, using Zeppelins and large Gotha biplanes, had caused near panic in Lloyd George's government. He had appointed the South African General, Jan Smuts, to examine the problem and his conclusion was that the only real response was a larger counter-offensive. Little had been done by the time the First World War ended, which meant that British theories of strategic bombing remained untested. Nevertheless, they did become the new orthodoxy in Britain between the wars, even though such exercises as were held tended to show that there was a significant gap between expectations and results.

The installation of the Nazi government in 1933 eventually led to a policy of rearmament in Britain, in response to Hitler's actions in Germany, but whereas Germany concentrated mainly on creating a war machine superbly integrated at the tactical and operational levels, the RAF began to build up a strategic bomber fleet.

When the war broke out in 1939, Britain did not immediately launch the strategic air offensive for fear of provoking a similar response from the larger and apparently more effective Luftwaffe, but what both the British and Germans did learn was that long-range bombing by day without a substantial fighter escort was tantamount to suicide. Consequently both sides switched to night bombing, even though studies showed that accuracy was actually so impaired that the smallest target that could be attacked with any expectation of doing any damage was a medium-sized town. As the threat of German invasion receded and then lifted during the course of 1941, and with the largest part of German resources directed towards the east and the campaign in Russia, Britain was able to take the initiative in the air war and bring to bear an increasingly effective strategic air arm against Germany.

At this point, with the realization that night bombing accuracy was poor, the British government was faced with a choice: either to hold back until some means of improving accuracy was found; or to accept a policy of 'area bombing' in the hope that the moral effect on German civilians would make up for the lack of physical effect on military and industrial targets. This was an idea that was taken up with some enthusiasm by the new commander of the RAF's Bomber Command, Air Chief Marshal Sir Arthur Harris, at the beginning of 1942. Harris had witnessed large areas of London burning during the Blitz in 1940 and was convinced that with the forces at his disposal he could do better. At the end of May 1942 he gathered together almost the entire strength of Bomber Command and launched over 1,000 aircraft against Cologne, causing massive damage. Thereafter, using his large fleet of heavy bombers (British bombers could carry much larger loads than their counterparts in other air forces; the Lancaster, Britain's best bomber, could carry more than twice the load of an American Flying Fortress, and more than three times the load of a German Heinkel), Harris began a systematic campaign designed to destroy Germany's industrial base and civilian morale.

In contrast to the British, the US Air Force remained convinced that it would be possible to bomb Germany by day, relying on the heavy armament carried by their bombers and hoping to be able to match the damage caused by the larger British forces through superior accuracy, but this turned out not to be the case. American attacks on targets in Germany began in January 1943, but losses were heavy and a series of disastrous operations culminated in an attack on the German ball-bearing factories at Schweinfurt in October 1943 during which, out of a force of 291 aircraft, over a hundred were severely damaged and sixty shot down.

The Americans came up with their own solution to this problem: instead of switching to night attacks, like the British, they decided to use escort fighters to defend the bomber force. They were able to do this because of a surprise discovery. In 1942 the RAF had issued a specification for a second-generation fighter to supplement the Spitfire and had eventually ordered an American development called the Mustang, which had by then been rejected by the US Air Force as underpowered. The RAF had asked North American, the Mustang's manufacturer, to fit the plane with the Rolls-Royce-designed Merlin engine, and it was found that with this modification it became a very superior, high-performance aircraft, which, when additionally fitted with long-range disposable fuel tanks, was perfectly capable of reaching targets throughout Germany and defeating all current German fighters on their own terms.

The combination of escort fighters and large, heavily armed bomber formations was more than a match for the Luftwaffe by day. A change of tactics by the RAF heavy bomber force at night was to prove devastating. The RAF had realized that the best solution to the night accuracy problem was to develop an elite Pathfinder force which would combine the best crews with the most modern electronic navigational aids in order to mark targets and then guide subsequent waves of bombers on to them with an accuracy that proved superior to anything the US Air Force could achieve by day. In February 1944 the RAF and US Air Force mounted 'Big Week': a combined, co-ordinated attack against the German aircraft industry, during which the British and Americans dropped some 20,000 tons of high explosives and managed to take out just under 34 per cent of German fighter aircraft available to defend Germany. In effect the Western allies had, by March 1944, severely undermined the Luftwaffe's ability to defend Germany; they had achieved a huge quantitative and qualitative superiority over the Luftwaffe; and the air war was all but won.

Ironically, as the Luftwaffe slid inexorably towards defeat, it was on the verge of introducing a series of aircraft types an entire generation ahead of anything that the Allies were capable of fielding, which, had they been available – as they might easily have been – only a year earlier, could well have brought about a strikingly different outcome to the war in the air.

In the mid-1930s it was becoming increasingly apparent to aeronautical engineers and designers that the piston-engined, propeller-driven fighter aircraft was approaching the limits of development. Having evolved from the fragile structures of the First World War, built mostly from canvas stretched over wooden spars, with temperamental engines barely capable of pushing the speed past 150mph, most fighters were, by the mid-1930s, assuming the final form of their type: all-metal, low-wing monoplanes with enclosed cockpits and engines capable of pulling them along at speeds above 350mph. Around this time, forward-thinking designers began to ponder what would come next. A solution which occurred to several, including an RAF officer in Britain called Frank Whittle, and a research assistant at the University of Göttingen called Hans-Joachim von Ohain, was to build an aircraft around a gas-turbine jet engine.

Von Ohain sought and received backing from the Heinkel aircraft company to develop and build an experimental engine, and ran it on the test bed in September 1937, some four months after Whittle had achieved a similar result in Britain. But while Whittle's engine engaged the attention of the British Air Ministry, this did not formally issue contracts to develop it for a further two years. By contrast, von Ohain received immediate encouragement and funds to build a practical aircraft engine, and on 27 August 1939, three days before German troops crossed the border to begin their attack on Poland and thus start the Second World War, the first practical jet aircraft, the Heinkel He178, took off on its maiden flight from Marienehe airfield.

More flights were made, and in November 1939 the He178 was first demonstrated for the head of the Luftwaffe's technical office, General Ernst Udet, a First World War fighter ace and crony of Hermann Göring. Udet was not impressed by what he saw. Although he had been responsible for arming the Luftwaffe with a superb range of tactical aircraft, he did not see the need for jet-powered fighters and as the top speed of the He178 was then only some 373mph – about the same as the faster propeller-driven aircraft – it did not seem to offer a great deal of improvement over existing types. Undeterred, von Ohain and his team continued with development and in March 1941 were ready to demonstrate the first practical jet fighter, the He280.

The He280 took off for the first time on 30 March 1941, nearly six weeks before the British Gloster E.28/39, Whittle's first jet-powered aircraft. The He280 undertook its maiden flight successfully, with no engine cowlings, in order to minimize a fire hazard which had shown up in ground tests. The engines used were two Heinkel-Hirth HeS8 turbojets. The wings were straight and mid-set, with the engines slung underneath about one-third of the way from the fuselage. A long, pointed nose led back to the single-seat pressurized cabin, and the high-set tailplane carried twin fins.

A few days later it was demonstrated for Udet, but again he was unimpressed, even though the aircraft had achieved a speed of 485mph in level flight at 20,000 feet – a good 100mph faster than any piston-engined fighter.

There were several unique features about the He280. It was the first twin-jet aircraft, the first jet aircraft to go beyond the prototype stage and it was fitted with the world's first ejection seat. The reason for this was highly practical: at the speeds it was capable of achieving, escape for the pilot becomes problematic because the slipstream is likely to tangle the pilot up in the airframe unless he can be thrown well clear.

Eventually eight He280s were built and one of them was pitted against a Focke-Wulf 190 piston-engined fighter in a mock battle late in 1941 to prove the jet's superiority. The jet won the duel easily, but even so there was little interest in it and the project was cancelled.

Part of the problem for the early German jets was the personal antipathy which existed between Udet and Ernst Heinkel, and even while von Ohain and Heinkel's company were pioneering research into jet aircraft and engines, the Luftwaffe was inviting its commercial rivals to develop their own airframes to accommodate jets. This was the origin of the aircraft which eventually became the Messerschmitt 262, which was actually the first practical jet fighter.

Preliminary design work on the Me262 was actually complete by June 1939, but the first prototype Me262 airframe was finished well before its jet engines. The airframe design was produced by Dr Woldemar Voigt, who came up with a low-wing monoplane with a slight sweep on the wing leading edge. The first prototype was finished in April 1941, but at that point the engines were still being bench-tested. BMW, which had been tasked to produce them, was initially unable to get its engines to deliver enough thrust, and for this reason the prototype made its first flight on 18 April with a conventional Junkers Jumo 210 G piston engine mounted in the nose.

By November 1941 the BMW 003 engines were ready for installation into the 262 airframe. On 25 March 1942 an Me262 prototype took off under the power of a Jumo 210 piston engine and two of the jet engines. The inclusion of the piston engine proved to be a wise decision because, as the aircraft approached take-off speed, both jet engines failed one after the other owing to compressor-blade failures, leaving the test pilot, Fritz Wendel, to land the aircraft solely under the power of the 'old' propeller.

Further development was continued, but with use of the new Junkers Jumo 004 Turbojet engine. This was more reliable, producing 2,200lb of thrust, and enabled Wendel to take off for the first time solely by jet power on 18 July 1942. He reported: 'My turbojets ran like clockwork and it was a sheer pleasure to fly this new machine. Indeed, seldom have I been so enthusiastic during a first flight with a new aircraft as with the Me262.'

Following this, the chief test pilot from the Luftwaffe's experimental station at Rechlin was invited to fly the aircraft and did so. Unfortunately he misunderstood his briefing – the Me262 was difficult to get into the air because of the lack of propeller slipstream over the control surfaces – and overshot the runway on take-off, leaving the aircraft damaged in a cornfield. Despite this, fifteen aircraft were ordered for evaluation, and in December 1942 it was decided to begin very limited production at the rate of twenty a month.

The real turning point for the Me262 came in April 1943

when General Adolf Galland, the chief of the Luftwaffe's fighter arm, flew the fourth prototype 262. Highly impressed, he reported to Göring that the plane would give German pilots a marked superiority over all Allied propeller-driven aircraft, and at a conference in June it was agreed to put the Me262 into production. Despite this decision, made three months after the first prototype of the first British jet fighter, the Gloster Meteor, had flown for the first time, the Me262 still faced serious difficulties, but the problem now was not technical but political.

The Me262 was demonstrated to Hitler at his headquarters in East Prussia for the first time in November 1943, and he immediately agreed to its mass production, but as a bomber. This bizarre decision, contrary to all the advice he had received, was made because Hitler was increasingly anxious about the prospect of an Allied invasion of France. In his mind's eye he envisaged the jets appearing above the invasion beaches at the critical moment, immune from ground fire because of their high speed, breaking up the Allied ground forces with cannon fire, rockets and bombs.

The reality was somewhat different. The Me262 had no provision whatever to carry bombs: it was a pure fighter, capable of 540mph at 20,000 feet, more than 100mph faster than the P51 Mustang, but Hitler could not be persuaded to change his mind. With an external bomb load attached, the Me262 was no faster than the Allied propeller-driven fighters and suffered accordingly. Although the first operational unit of Me262s was formed in April 1944, with the dual function of training pilots and shooting down Allied air reconnaissance planes, fewer than 150 of the 1,433 Me262s that were built ever made it into combat use. Even after Hitler reversed his decision in November 1944, as the result of the steady increase in US daylight raids, and demanded that the Me262s be converted back to fighter use immediately, there were too few aircraft and too few qualified pilots to make much difference. Even so, the Me262 represented a considerable shock to Allied pilots who encountered them in the closing months of the war, as test pilot Eric Brown recalls:

> Towards the end of the war I should think those that were fighting in Europe were finding that the German Luftwaffe was equipped with aircraft which really were almost unassailable in the sense that they had the speed to escape from most of the piston-engined fighters that were around. But of course life's not just as simple as that. The jet aircraft which the Germans had had a problem: the engines at that time were very, very slow in their acceleration and consequently on take-off the acceleration was slow and the aircraft had a long haul before it got off, got its undercarriage up and got to the climb phase; during that period it was very vulnerable.
>
> Likewise in landing, because none of the early jets had dive brakes and in consequence they found it very difficult to slow up for the approach speed so they tended to have long, low creeper approaches. Again they were very vulnerable and when we twigged this, of course, we set up standing patrols over the airfields where they operated on with Mustangs and Thunderbolts etcetera. They suffered badly at that time: that's when most of the German jets were shot down, in these two phases, not so much in combat, and eventually they had to counter this by themselves sending up standing patrols of Focke-Wulf 190s over their airfields and that neutralized the situation a bit.
>
> But it was quite a shock, I think, for many to find that they were being left absolutely standing by the German jets, because the Messerschmitt 262, for example, had probably an advantage of at least 100mph over any contemporary piston-engined fighter.

But if the Me262 represented an evolutionary step from fast piston-engined, propeller-driven aircraft to the basic jet fighter, it was its tiny cousin, the Me163b Komet, which had so surprised American pilots on its first appearance in May 1944, that was the real revolution: an aircraft that pointed the way to supersonic flight and to men in space.

The origins of the Komet lay in the designs of Alexander Lippisch, who had for some years been experimenting with delta-wing gliders. Lippisch had experimented by putting a relatively low-powered rocket engine into one of his gliders, the model 194, and had managed to get it to fly at speeds of more than 300mph. To follow this Lippisch was given access to a Walter rocket motor which had originally been intended for an experimental fighter similar to the He178, to see whether it might be possible to build a rocket-powered interceptor fighter. In collaboration with the Messerschmitt company, Lippisch designed and built a small, unpowered delta-wing airframe to conduct early flight tests.

During one of these tests, in the summer of 1940, the pilot had been towed in the glider, now designated the Me163a, up to 16,000 feet and then released to glide back to the airfield at Augsburg, which General Udet happened to be visiting. The pilot put the glider into a steep dive, bringing its speed up to 400mph, before sweeping past Udet, pulling into a vertical climb and then returning for a perfect landing. Impressed by the display of aerobatics and the speed of the tiny aircraft, Udet asked what engine it had and was amazed to discover that it was unpowered. He immediately gave his authorization for further tests with a much higher priority.

The rocket-power trials took place at the research establishment at Peenemünde West in the summer of 1941, using a Walter HWK R11 203 giving a thrust of 1,653lb. The flying characteristics of the Komet were excellent, and with the amount of power available, Heini Dittmar, the test pilot, had no difficulty in exceeding the world speed record, which then stood at 469.22mph. On the fourth flight he took the Komet to 571.78mph, but the fuel ran out while the aircraft was still accelerating and he was forced to glide back to base. Even so, Lippisch's calculations suggested that the aircraft ought to be able to reach at least 1000kph (621.4mph) and to test this theory, Dittmar suggested that it be towed up to altitude so that it would not need to waste fuel in the take-off. As a result, on 2 October 1941 Dittmar was towed to 13,000 feet by a Messerschmitt 110. There he fired the rocket engine and, in level flight, the Komet accelerated to 1002kph (622.6mph), or Mach 0.84. It was now that he began to experience problems:

> My airspeed indicator was soon reading 910kph and kept on increasing, soon topping the 1000kph mark. Then the needle began to waver, there was sudden vibration in the elevons and the next moment the aircraft went into an uncontrollable dive, causing high negative 'G'.
>
> I immediately cut the rocket and for a few moments thought that I had really had it at last!
>
> Then, just as suddenly, the controls reacted normally again and I eased the aircraft out of its dive.

In fact what Dittmar had experienced had been a 'compressibility stall' caused by his aircraft flying close to the speed of sound. When this happens, parts of the airflow over the wings are often going faster than the speed of sound, and as a result, unpredictable turbulence, vibration and buffeting occur. This had been noted by pilots of propeller-driven aircraft in power dives, but never in level flight before, and in fact the swept-wing design was the best way of combating it. Even though the first British jet fighter, the Gloster Meteor, was operational in the late summer of 1944, it did not have a swept wing and suffered accordingly, according to Eric Brown:

> Yes, this of course was one of the great aerodynamic innovations in which we missed the boat really, because the Germans had been experimenting for a considerable time on the qualities of the swept wing to delay the compressibility stall in transonic flight and Professor Busemann, who was really the leading expert on this in Germany, actually gave an open conference on it in Rome in 1935 which was attended by many nations: France, Britain, the United States; and somehow or other, although many notes were taken etcetera, it wasn't followed up. So the Germans went ahead very much on their own in this area, and of course it paid them huge dividends because it gave them a tremendous advantage in the transonic region where at the end of the war much of the fighting was being done; and we were at a very serious disadvantage, so one must put it down as one of the great aerodynamic innovations.

Pleased with the results of the tests, Udet ordered seventy rocket-powered Me163b Komets as pre-production prototypes, with the intention of having a rocket interceptor unit operational by spring 1943. But it was not to be. In November 1941 Udet, a depressive alcoholic and narcotics abuser, shot himself. His successor, Field Marshal Erhard Milch, confirmed the order for the Komet but declined to give it as high a priority as Udet had, and the project began to drift. This was partly because Lippisch did not enjoy working with Messerschmitt but also because German military research was now strongly afflicted by short-termism.

It had become evident, by the time the German Sixth Army had become trapped at Stalingrad in the winter of 1942–3, that the war was by no means settled and that if Germany was indeed to win it, then considerably greater sacrifices would have to be made than hitherto, if the war was not already lost. In part this was sensible: it prioritized industrial production towards basic weaponry like tanks,

ABOVE: The 'Doctors' Trial' at Nuremberg, July 1947. Karl Brandt, (1st row, far left) Hitler's surgeon and one of the organizers of T-4, received a death sentence and was hanged on 2 June 1948.

ABOVE: The remains of Heisenberg's 'uranium burner' are discovered by the ALSOS mission in Haigerloch, May 1945. Had Heisenberg succeeded in making it go 'critical', its lack of shielding would have caused a nuclear catastrophe.

ABOVE: Cubes of uranium oxide, fuel for the uranium burner, dug up by ALSOS specialists from the field where they had been buried.

ABOVE: Werner Heisenberg (left), Max von Laue (centre) and Otto Hahn at Göttingen following their return from internment at Farm Hall, 1946.

ABOVE: Heisenberg teaching at Leipzig University in the 1930s.

ABOVE: Many of the key figures from both the German and Allied atomic bomb projects attended this nuclear physics seminar in Copenhagen in 1937. Bohr, Heisenberg, Pauli, and Meitner (1st row, 1st, 2nd, 3rd, and 5th left, respectively); Peierls (2nd row, 4th left), von Weizsäcker (3rd row, 2nd left); Oliphant, and Frisch (back row, 2nd and 3rd left, respectively).

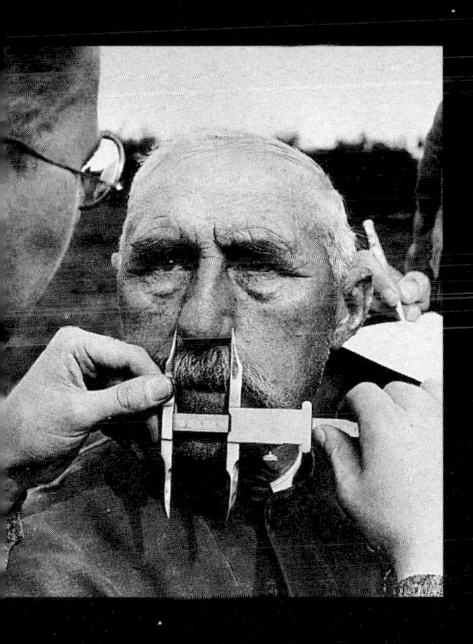

ABOVE: Eugenic anthropologists from Kiel University measuring the faces of villagers in northern Germany in 1932

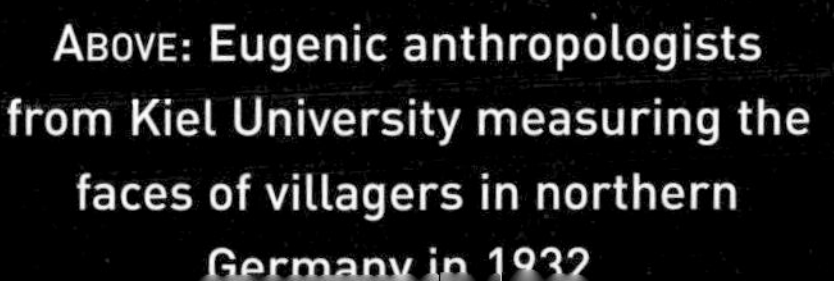

ABOVE: Doctor Ploetz, author of the key euthanasia text *Life Unworthy of Life* in 1931

Girls learning the characteristics of the inferior races

LEFT: With its slogan 'The Dreadful Legacy of a Drinker', this propaganda poster depicted the degenerative consequences of alcoholism.

LEFT: Blonde and athletic, this javelin thrower – featured on a 1934 postcard – epitomised the eugenic Aryan ideal.

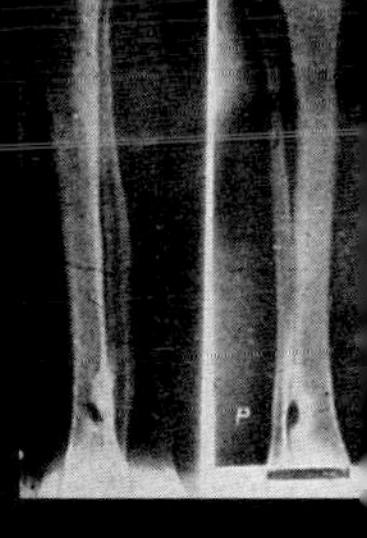

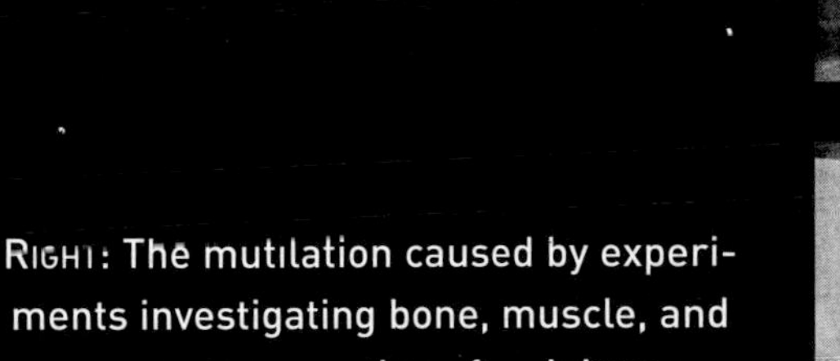

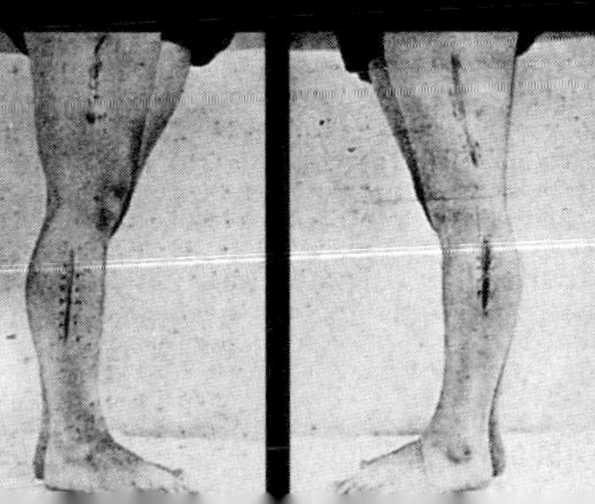

RIGHT: The mutilation caused by experiments investigating bone, muscle, and nerve regeneration after injury. The unwilling victims were wounded and deliberately infected with bacteria, often with fatal results.

Doctor Muench (inset), camp medical officer at Auschwitz Birkenau and organizer of the burning of human remains, as depicted by this photograph taken in secret.

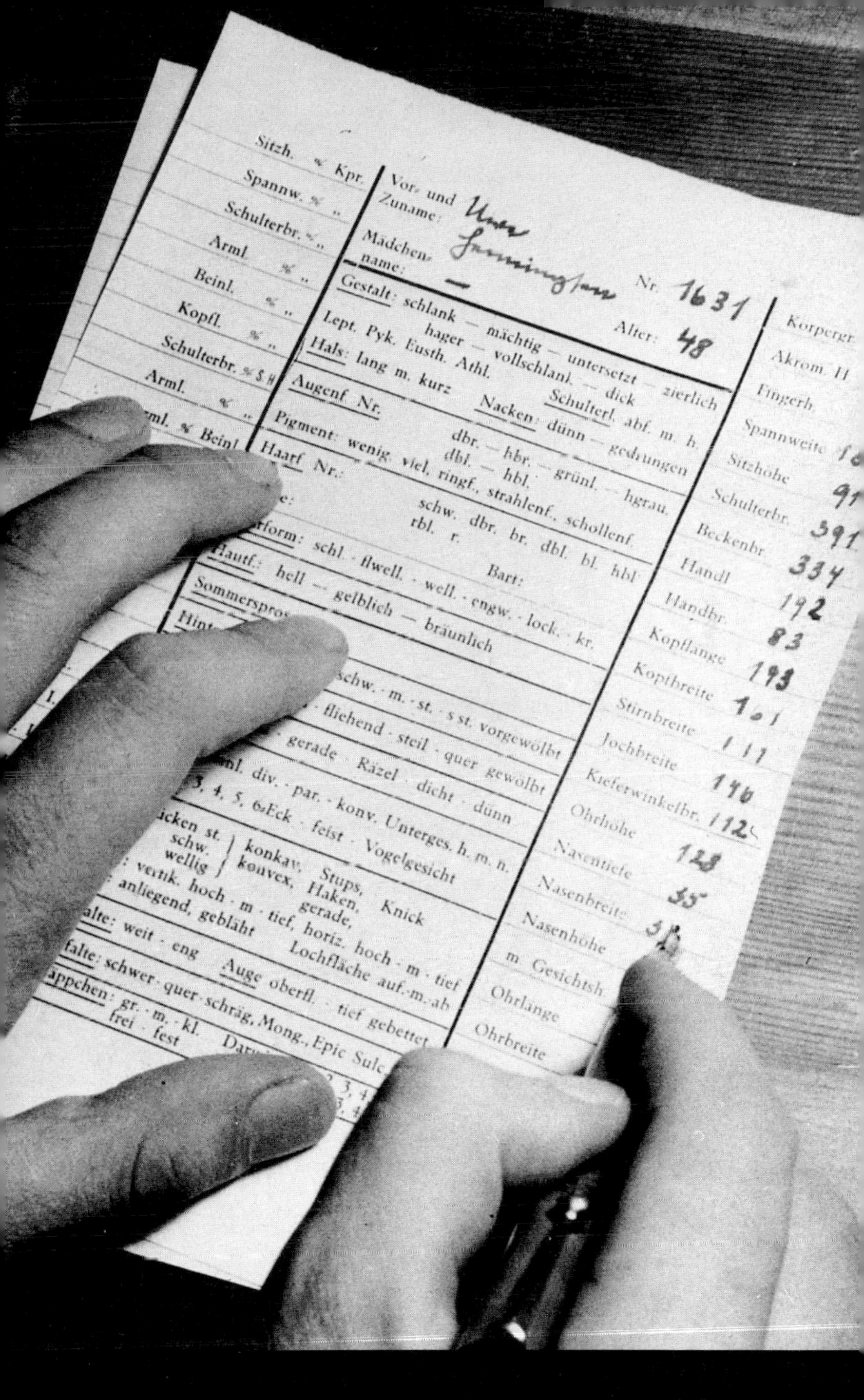

The eugenic questionaire used by anthropologists gathering data from German villagers, 1932

LEFT: A concentration camp inmate in a simulated parachute drop in the low-pressure chamber at Dachau, 1942.

BELOW: Doctor Sigmund Rascher (right) with a colleague and an experimental subject during the low temperature experiments at Dachau in 1942.

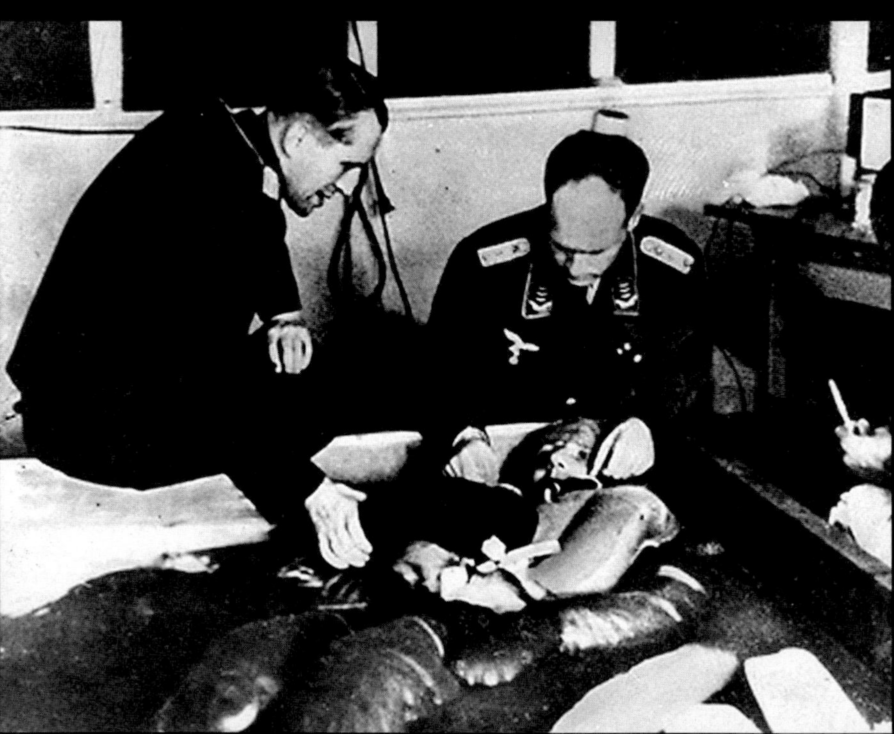

fighters, bombers, rifles and steel helmets; but at the same time, it left many potentially important projects hanging in mid-air. These then needed ministerial approval before they could continue.

Such approval for the Komet came in April 1943, when the Armaments Minister, Albert Speer, ordered that 1,000 aircraft should be produced to provide point air defence at strategic targets like the synthetic oil plant at Leuna near Leipzig. Point air defence is an unusual task for an aircraft rather than a gun or missile system, but the characteristics of the Komet made it ideal. The tactics adopted by the first operational Komet squadron were simple and direct: as enemy bombers approached, the Komets would launch and accelerate almost vertically at 440mph, taking a brisk five minutes and forty seconds to reach 37,000 feet. At that point the aircraft would level off and accelerate to about 560mph, directed by ground radar to a point some 3,000 feet above the US bomber formation, at which point they would descend from all points of the compass like a swarm. In theory the Komets would then have around two minutes of powered flight left to make their passes at the bombers before breaking of their attack and gliding back to base. The reality was considerably different.

The main problem that operational Komet units faced was the fuel. The Walter HWK 109-509A rocket motor gave 3,307lb of thrust for a weight of 366lb, but the fuel that it used caused difficulties. The rocket worked on the basis of mixing 'C Stoff' (57 per cent methyl alcohol, 30 per cent hydrazine hydrate and 13 per cent water) with 'T Stoff' (80 per cent hydrogen peroxide). Both of these fuels are dangerous enough on their own, but in combination they instantly decompose into high-temperature gases. As a result, even refuelling became an operation pregnant with risk. The two separate fuel bowsers, categorically marked 'C' and 'T', were not permitted within a kilometre of each other, and after each tank was filled everything connected with the operation was hosed down with water to ensure as little danger as possible. In training films made for Luftwaffe ground crew at the time, a drop of T Stoff is shown being dropped into a saucer of C Stoff: the result is a detonation with the force of a rifle shot. There was another problem associated with T Stoff, as Komet pilot Mano Ziegler explained:

> ...Joschi's aircraft dropped like a stone, hit the airfield at an angle and skidded along the ground for some fifty metres before coming to a standstill. All this had taken place at least two kilometres from our flight line where we had been standing, and we ran as fast as our legs would carry us.
>
> ...surely it could not be too bad. At least there had been no fire and no explosion. The fire engine and ambulance reached Joschi's machine within a minute of the crash, but our Joschi was no more. The T Stoff from fractured fuel lines had seeped into the cockpit and poor Joschi, probably unconscious as a result of hitting his head on the instrument panel, had been dissolved alive...

When the time came to start up the aircraft, tension ran high: first the engine and jet pipe needed to be thoroughly doused with water from a high-pressure hose to ensure that there were no lingering fuel residues that could cause an explosion, and then an electric pump would introduce a small but precise quantity of calcium permanganate and potassium chromate to the T Stoff. This would start a chemical reaction, delivering high-pressure steam to the main pumps, which would begin the mixing of the two main fuel components. In May 1945 Eric Brown was the first Allied pilot to fly the Komet:

> From January 1944 I was the chief naval test pilot at the Royal Aircraft Establishment at Farnborough and I was also in the aerodynamics flight

and the high-speed flight. And in 1945, because I was German-speaking, it was decided to make me commanding officer of the captured enemy aircraft flight, which meant that at the end of the war we were to go into Germany and get hold of as many experimental aircraft – or modern advanced aircraft – as we could before they had a chance to destroy them.

Well, I went in about two or three days before the capitulation and we were really following up our troops as they advanced northwards and we were asking them not to destroy the particular kind of aircraft we hoped to get. Of course the Germans themselves had destroyed a huge number of aircraft, but strangely enough they didn't destroy the ones we were interested in, which were the modern jets, and I believe this was because of a feeling of pride in what they had in these very advanced aircraft. So we were very fortunate and picked up quite a large number of them.

That was just after the capitulation and a clamp had just been put on by the C-in-C Germany that we should not be flying around in German aircraft. Except this did not apply to the RAE test pilots, who were given virtually a carte blanche to fly them.

It was quite obvious that we were not going to attempt to fly the 163 in England, so I could see I was only going to have one fleeting chance of getting a flight in one, which I did at Husum in northern Germany. And I did this with the cooperation – and rather unwilling cooperation – of the German ground crew because the whole problem was dealing with the very volatile fuels which were used for the rocket motor of the 163. I only had one flight and I was very glad I did because it was quite an extraordinary sensation; this was just an aircraft that was in the category of the North American X-15: something quite unique and very exciting indeed and in many ways very hazardous.

In fact Brown had made a number of familiarization flights in the unpowered glider version before risking the Me163b. Nevertheless, it was a prospect that was daunting even for the trained Luftwaffe pilots who had flown the aircraft operationally, few of whom had taken to the air in the rocket version more than a dozen or so times:

> Once you're in the cockpit and you've got yourself started up, the aircraft is on a small two-wheeled trolley and that trolley is attached to the landing skid under the aircraft, which is extended at this stage for the trolley to be hooked on to it.
>
> As you start off on the take-off run under rocket power, it's a very, very fast acceleration, and you come to quite a critical moment when you lift off, because the height at which you jettison the undercarriage – the little trolley – is quite critical. If you jettison it too low, it could bounce back up and puncture your fuel tanks; if you jettison it too high you get a really good bounce and it could actually come up and hook on to the front of the skid... This actually happened to the woman test pilot Hanna Reitsch.
>
> But provided you've got the right height, which is about fifteen feet, and you get away, you retract the undercarriage and what is a normal-angle take-off then becomes a startling zoom climb where the attitude of the aircraft is about forty-five degrees going uphill. And one must realize that at a time in fighter technology when the average contemporary piston-engined fighter had a rate of climb of about 3,500 feet a minute, this aircraft had an initial rate of climb of 16,000 feet a minute!
>
> So you were going up like a lift and it was very difficult to see the horizon at this angle: you were virtually on your back in the cockpit, so you had to rely a lot on instruments until you got up to the height at which you wanted to cut the power, because ... tactically the aircraft was supposed to zoom up above the B-17 bomber streams. The pilot would cut the power and ... had enough speed in his dive with his swept-wing arrangement, he could come down on to the B-17s and at a range of about ... 600 yards – because he was armed with two 30-millimetre cannon and they're not very accurate above 600 yards

– you can open fire ... and you have to break away to avoid collision at about 200 yards, so you really have only two seconds' firing time. But two 30-millimetre cannon are very destructive, so this was adequate.

And once you broke away – your stopwatch was very important in all this because when you cut the throttle, stopped the rocket motor, two minutes must elapse before you could relight, but if you followed that routine you could get in about six passes like this and zoom back up, come down again ...Of course it was a totally fair-weather machine because you had to be able to see the airfield where you were going to land because it was a glide landing.

But the most daunting aspect of the Me163b was its sheer speed and the odd characteristics it displayed when flying at its limits:

In the 163 ... not during my flight in Germany but when we had more time to get into the real phased testing of it at Farnborough, we took it up to its Mach limit of .84. In transonic flight you have a critical Mach number at which ... if you go beyond that you will lose control, and in the case of the Me163 that was .84 times the speed of sound, and if you went beyond it in this case the reaction was really violent: the aircraft nosed steeply down into what we used to call a graveyard dive and there really was no way out if you let it get much beyond .84, it was very dangerous in this respect.

This was a very advanced concept indeed in three areas: it had swept wings, it was tailless and it had rocket power. Three new areas for us and they were all in the one plane.

The aeroplane as such flew very, very nicely indeed in spite of being tailless – when I say tailless it had a vertical tail but it had no horizontal tail – the elevators were really incorporated in the aileron so there were elevons on both wings. In spite of this the stability and control of the aircraft were first class, except if you got beyond this limiting Mach number, this critical Mach number.

But in addition to the completely novel experience of controlling a plane flying in this manner, the Komet presented basic physical difficulties as well:

> ... there was a lot of G experienced and a good kick in your back from the rocket thrust, not excessive but quite a lot and of course the unusual altitude gave you problems of sighting but I had no pressurization problems, the aircraft was not pressurized, the cockpit was an unpressurized cockpit, but I'd done a lot of flying up to 40,000 to 45,000 feet in both pressurized and non-pressurized aircraft so I didn't find any particular problem in that respect... mind, I only took the aircraft up to about 38,000 feet. I think ... theoretically it was capable of going to 60,000 feet. I think if you'd gone up much beyond 40,000 feet you would have had problems because it was a non-pressurized cabin.
>
> ...with an unpressurized cabin you definitely feel that your limit is around 35,000 feet because I've been beyond that a lot but in pressurized cabins, and in the unpressurized cabin you just had that innate feeling that you are getting near the edge of maybe losing consciousness if you go much higher and getting into a serious cockpit environmental problem.
>
> Well, psychologically, of course, you do feel very, very claustrophobic. It's a small cockpit, you're locked in there and you know this aeroplane is capable of about 600 miles an hour and you're in an aeroplane which has no ejection seat. The maximum limiting bail-out speed to get rid of the hood and get out was 250 miles an hour: this in an aeroplane that flies at 600 miles an hour and lands at 125, so you are really in a tin coffin for a large part of the time. I think you're conscious of this, not to a great degree but it's in the back of your mind.
>
> Then, as I say, at high altitude you are very conscious of the fact that you mustn't go too far in height because you're in an unpressurized cabin and then the landing problem: you know it's a glide landing, you've got to get it right; you've got to find the airfield, get it right first time.
>
> And above all you have to make sure that you haven't got a drop of fuel

> in that aeroplane because the fuels that were used for the rocket were highly volatile and any amount left in the tanks could explode if you make a heavy landing, and most of the landings were heavy-ish because you're landing on a skid. So you had to make sure you jettisoned [all fuel].
>
> And many, many of the pilots were killed in landing or broke their backs. Now, the reason they broke their backs was the hydraulic skid which you landed on was put down: you pressed a lever down and the skid lowered but once you got the light showing it was fully lowered, you had to return the lever to neutral, otherwise the skid was left under hydraulic pressure and would be rigid, whereas once you put the lever back you released some of the pressure and it could absorb the touch-down. So many pilots coming back under stress or maybe stress of combat, stress of finding the airfield, forgot to put it back in neutral. So broken backs abounded amongst 163 pilots.

Brown never flew the Komet in combat, of course, but he was nonetheless sceptical of its combat potential, despite its startling innovation:

> Well, to put it in perspective, I think that it cost the Germans a lot more aircraft and men than it ever did the Allies. It was effective if somebody was flying it to the right combat pattern, and with two 30-millimetre guns it was a very powerful armament. But you had to get yourself in the right position and ... under full power your total flying time – that is when the power was being used – was 120 seconds. So you had to be really on the ball. I think the majority of 163 pilots were under so much stress flying the aeroplane that this reduced their combat efficiency tremendously.

As a highly experienced test pilot, Brown regarded the physiological aspects of flying the Komet as relatively routine, but the evidence is that this view was not shared by the Luftwaffe, which appears to have concentrated on the physical preparation of Komet pilots as much if not more than on flight training. In 1944 Kurt Schiebeler was selected to join the early stages of the operational Komet deployment:

> I was with Fighter Squadron 54 in Russia, and each squadron had to supply one pilot who had some front-line experience. Since our squadron had already been heavily decimated and I was just one of those who had some experience at the front, and was also young and in the best of health – the commander of my squadron was the famous fighter pilot Nowotny – and he was the one who suggested me for this business.
>
> I had no idea where it would all lead or what it was for, you see, [so] I tried to persuade Novotny to keep me on there. I would have preferred to stay in Russia, because I was afraid I would be sent for high-altitude combat: we just had an idea it was to do with high-level combat, you know; flying at high altitudes. But Nowotny didn't know anything either and I had to go, because there was nobody else to do it.
>
> And so I followed a very roundabout route: I then flew in an Me109 from Smolensk towards Germany, with a few stops in between, and then I reached Bad Zwischenahn [the base for the Komet training unit] and reported there and it turned out that practically the whole business was still in its very early stages: there were actually a couple of aircraft there, but nothing was ready to fly or fit to use at all, you see.
>
> I was told: 'Just go up to the top of the mountain for four weeks, you'll be taking some mountain air.'
>
> I thought, oh, so this war could be quite good fun: I certainly hadn't expected that. And I thought, well, if that's the case, you will gradually get through this war. I then went for this high-altitude cure, but first I went to a centre in Munich for an altitude test, and in this test I got to a height of 8,000 metres before I became unconscious.
>
> And then I went to the top of the mountain, came down again after four weeks and then the test was repeated ... we had to keep writing: 'High-

altitude flight fitness investigation centre': all one word in German and a tough one to write out, too.

I kept writing it for quite a while and then got to a height of 11,000 metres, which in itself seems quite hard to credit, really. Normally people would be dead by then, from 5,000 metres upwards people usually die... their lungs can't cope any more.

And because of this training I got as far as 11,000 metres and then I became unconscious as well. All this was under the supervision of a doctor. The doctor then gave me an oxygen douche and that sorted me out, just like that.

Returning from the mountains, Schiebeler was taught the rudiments of flying the Komet in the glider version, as well as the low-powered ME163a, before being allowed to take up the 163b for the first time:

... none of us had any idea how people perform in a steep angle of climb at high speed. So we then did training beforehand, and we did this every day for twenty minutes in a vacuum chamber, where we simulated this angle of climb. But it was impossible to simulate it exactly as it would be in reality, so there was nothing else for it but simply to test the feasibility of the idea in practice.

And so we then tried it out; the first people to do this were Rudolf Opitz and Wolfgang Späte, and later there was also Hanna Reitsch, a famous woman flyer, who also trained with us, you see.

And then I had the experience of the first take-off, when I ... did one take-off with a quarter load, in an aircraft with the fuel tanks one quarter full. With this load, we reached an altitude of roughly 3,000 metres or 10,000 feet.

And then we already began to realize that during the steep climb we had to take really deep breaths. We were short of breath, because the body first had to adjust to the external pressure. The external pressure falls steeply and the pressure within the body rises, so that it was important for us not to eat anything that would cause any flatulence at all.

> And that was all very well as far as it went, but for the second take-off the aircraft fuel tanks were half full and then we got as far as about 20,000 feet and then things started to become a bit more difficult. I mean, I can well recall that then I had to, from a height of 3,000 metres, had to really force my mouth right open all the time and take deep breaths, in order to balance the pressures.
>
> But somehow, as a young man you manage to get through it, [but] whether it would have continued to be all right for very long, I don't know. We had no pressure suits and no pressurized cockpits either, because producing these parts was only a secondary priority.

Preparation for flying the Komet at high altitudes extended to the foods that the pilots were given: 'The meals consisted mainly of milk products and also meat and we had a special quartermaster with us who had to obtain our provisions, and a special cook who was responsible for us. Under no circumstances were pulses such as peas, beans and lentils and so on to be used, because then we just wouldn't have been able to stand it: then you get very severe flatulence.'

The effect of flatulence in a low-pressure environment would have been potentially disastrous. In theory, gases in the stomach and alimentary canal would have expanded as the pressure in the cockpit dropped, to a point where:

> The worst that could have happened is that our lungs could have burst, but nothing like that happened to us ... that was no problem for us at all.
>
> It was mainly due to the fact that we were constantly being trained for this ... In one test I actually got up to 11,000 metres: that's about 30,000, nearly 40,000 feet, you see ... well, no one believes that, when I tell them that nowadays ... because normally people certainly can't withstand that at all.

Schiebeler is one of the few men who took the Komet into combat against the US Air Force, and shot down an American bomber:

The first flight I ever did involving a B17 ... I nearly flew into it.

I mean, the approach speed was so colossal and we had to look through the sights if we were going to hit the enemy ... we were afraid we were about to fly right into the bomber and that made it quite impossible to take a properly targeted shot.

...we had to climb back above the aircraft by a few hundred metres and then make quite a normal approach ... in order to shoot it down, so it became clear that our whole mission as interceptors was ... doomed to failure from the beginning.

In the first place, the people that came to us were young, they had only learned to fly at training school, but they had no idea at all about shooting, and in a fast-flying aircraft, hitting an enemy with a cannon is extremely difficult, because of the approach speed ... there isn't enough time to aim.

With a slow-flying aircraft, to start with, the enemy is flying at a speed, let's say, of 300 kilometres and I approach him at a speed of 500 kilometres. This leaves a difference of 200 kilometres between us and that's not a lot: you can actually do something in that case.

But when you are approaching the enemy at 800 kilometres or 900 kilometres, the time available is so short that you have almost no time to hit the enemy ... This is why our bunch only managed to shoot down just a couple of aircraft.

I myself only shot down one, directly in front of the assembled team, which in my view was more down to luck than good management. Then I shot down a formation leader ... they always came over twelve aircraft together. I took out the leader and the others started flying backwards and forwards all over the place, they obviously had no idea at all where they were and had quite lost their bearings, and as a result a second plane was shot down by the flak.

By this time I had landed again, when down came – I still remember

> this – an American with a parachute and he was already half burned but was still alive. He came down right next to me and said: 'My name is Gottfried Krause from America!'
>
> Well, I was totally shaken and I started to think, whatever sort of a war is this? What's the point of it all, eh?

The accounts by Brown and Schiebeler of flying this revolutionary aircraft only really hint at one of the darkest secrets connected with its development. From an early stage the Luftwaffe had recognized that an aircraft capable of climbing to 40,000 feet in a couple of minutes was likely to exert extreme physical pressures on its pilot. Concerns about the effects of flying in an unpressurized aircraft at high altitudes were matched by worry about what effects might be caused by jumping out of them: a high-altitude escape from an aircraft without oxygen is fraught with danger because of the low temperature at which it must take place and the lack of oxygen. If a pilot jumps from a plane at, say 30,000 feet, he is faced with the choice of opening his parachute immediately, almost certainly becoming unconscious and risking frostbite on a slow descent; or free-falling to a level where the air temperature will be higher but he is exposed to the risks of rapid changes in pressure.

Trained pilots were scarce in Germany in the latter part of the war and every effort needed to be made to maximize their chances of survival and their operational lifespan. With eugenics and euthanasia lifting normal moral restraints from many doctors and with their enthusiastic acceptance of the Nazi movement, an obvious but hitherto untapped source of experimental data was discovered: human beings.

CHAPTER FOUR

the price of technology

The ability of the Luftwaffe to send pilots hurtling through the sky at close to the speed of sound, at the highest possible altitude and in unpressurized aircraft was not only based on their extraordinary technology. Without an understanding of the limitations of human physiology, that might well have turned out to be worthless, but such knowledge could only be gained by tests on humans, and in the ordinary course of events few were likely to volunteer to take part in such dangerous experimentation. But in the Third Reich this did not necessarily matter.

In order to investigate the effects of low pressure and high altitude on the human body, a doctor named Sigmund Rascher began a series of experiments to test to destruction the capacity of the human body to withstand these extremes. The subjects that he used were live humans.

Rascher was a thirty-year-old physician working in a junior job at a Munich hospital when the war broke out and he was conscripted as a medical officer into the Luftwaffe. It appears that Rascher's wife, Karoline, had a personal connection with Himmler: she had worked as a secretary to him at one time and may even have had a sexual affair with him, and Rascher himself was a member of the Allgemeine-SS, the black-shirted part-time political strong-arm squad. At the beginning of 1942 Rascher was stationed in or near Munich and took the opportunity to approach Himmler to obtain permission to begin a series of high-altitude experiments using inmates from the Dachau concentration camp. How far this was on Rascher's own initiative or whether he was acting on the instructions of the Institute of Aviation Medicine in Berlin remains unclear. What is certain is that Rascher was provided with an assistant to help him with the experiments and a special low-pressure chamber which was brought to Dachau from Berlin, and given access to as many prisoners as he needed. The experiments began in April 1942 and a report exists of the effects of one of the first of these on an unnamed thirty-seven-year-old Jew:

> The third experiment of this type took such an extraordinary course that I called an SS physician of the camp as a witness, since I had worked on these experiments all by myself. It was a continuous experiment without oxygen at a height of 12 kilometres on a 37-year old Jew in good general condition. Breathing continued up to 30 minutes. After 4 minutes the experimental subject began to perspire, and wiggle his head; after 5 minutes cramps occurred; between 6 and 10 minutes breathing increased in speed and the experimental subject became unconscious; from 11 to 30 minutes breathing slowed down to three breaths per minute, finally stopping altogether.
>
> Severest cyanosis developed in between [each breath] and foam appeared at the mouth.
>
> At 5 minute intervals electrocardiograms from three leads were written. After breathing had stopped the ECG was continuously written until the action of the heart had come to a complete standstill. About half an hour after breathing had stopped, dissection was started...
>
> When the cavity of the chest was opened the pericardium was filled tightly (heart tamponade). Upon opening the pericardium, 80cc of clear yellowish liquid gushed forth. The moment the tamponade had stopped, the right auricle of the heart began to beat heavily, at first at the rate of 60 actions per minute, then progressively slower. Twenty minutes after the pericardium had been opened, the right auricle was stopped by puncturing it. For about 15 minutes, a thin stream of blood spurted forth. Thereafter, clogging of the puncture wound in the auricle by coagulation of the blood and renewed acceleration of the action of the right auricle occurred.

> One hour after breathing had stopped, the spinal marrow was completely severed and the brain removed. Thereupon, the action of the auricle of the heart stopped for 40 seconds. It then renewed its action, coming to a complete standstill 8 minutes later. A heavy subarachnoid oedema was found in the brain. In the veins and arteries of the brain, a considerable quantity of air was discovered. Furthermore, the blood vessels in the heart and liver were enormously obstructed by embolism.

It is significant that a good proportion of the 'post-mortem' section of the report dwells on the effects of the damage that Rascher had inflicted during his dissection. Rascher's qualification to conduct medical research was dubious at best, and he is believed to have faked evidence to obtain the 'correct' results during pre-war cancer research. Later experiments describe simulated parachute descents. In one case a former delicatessen worker was given an oxygen mask and raised to a simulated height of 47,000 feet, at which point the mask was taken away and free-fall simulated. His reactions were described in detail in Rascher's report: 'spasmodic convulsions', 'agonal convulsive breathing', 'clonic conclusions', 'groaning', 'yells aloud', 'convulses arms and legs', 'grimaces, bites his tongue', 'does not respond to speech', 'gives the impression of someone who is completely out of his mind'.

Altogether around a hundred inmates of Dachau were killed during the high-altitude experiments, which came to a conclusion in August 1942, and at all times the Institute of Aviation Medicine was kept informed of Rascher's results.

Rascher's second series of human experiments began almost as soon as the low-pressure series ended. This experiment was to test the effects of extreme low temperatures on the human body, and to investigate how to re-warm patients who had been subjected to extreme chilling. The utility of these related to the problem of aircrew being downed in the sea, as Frank Golden, who studied the same problem in Britain in the 1960s, recalls:

> Just prior to the war in the medical literature there was quite a controversy going on about how you should re-warm people who were cold: some people said they should be rapidly re-warmed, and then another journal said no, you shouldn't rapidly re-warm them, you should slowly re-warm them and there were case studies proving each side of the argument. So nobody quite knew what to do.
>
> And then in the early years of the war, in the Battle of Britain time, many Luftwaffe pilots and our own Air Force pilots – and indeed American Air Force pilots later on – were being rescued from the North Sea, and it was found that many of these people were being rescued alive, albeit very cold, but were dying during rescue or shortly afterwards. So it was in an effort to try and resolve this question as to whether these people should be rapidly re-warmed or slowly re-warmed that I think triggered these experiments.
>
> About the same time there was some work being done by Professor Weltz in the Institute of Luftwaffe Medicine in Munich, where he showed some animal work. He in fact was interested in this rapid and slow re-warming story as well, and in order to try and find out what killed people with rapid re-warming he cooled some guinea pigs and then he rapidly re-warmed them intending to do some post-mortem work on the dead animals. To his surprise he found that most of the animals survived, so he did another experiment and he put them into even warmer water and even more survived.
>
> So that suggested to him that rapid re-warming was the thing to do. He then moved up to large pigs – or some colleagues of his did similar experiments – and again found that this rapid re-warming seemed to

work; so then [Rascher] did the next jump and went and used human beings as subjects.

That's the work ... that was done in Dachau and I think that's why it was done, to try and find whether you should rapidly re-warm somebody or slowly re-warm them, and if you're rapidly re-warming them what's the best way to do it: should it be by total immersion in a hot bath or should it be by body-to-body heat contact, [or by] a range of other procedures in between.

And the conclusion they came to was that the best thing to do was to put them into a hot bath which was right for their particular set of circumstances: it worked.

Rascher's experiments involved placing his subjects, normally dressed in Luftwaffe flying clothing, in a tank of iced water for hours on end. Golden describes the physiological effects that this causes:

When they first enter the water it's incredibly painful: I'm talking about cold water, very cold water, particularly anything below 10 degrees Centigrade is very cold and it gets particularly painful below 5, so they get this particularly painful experience: stinging, burning pain which then triggers a series of responses in the body, the most common of which are rapid uncontrollable breathing, their heart rate shoots up and their blood pressure shoots up.

Now those three responses will have an adverse effect on the person's survivability in the water. The increase in heart rate and the increase in blood pressure may to somebody who's already suffering from heart disease – high blood pressure, for example, or coronary artery disease – it may trigger a fatal heart attack or a stroke or something like that, and the uncontrollable breathing will make it impossible to hold your breath to escape underwater from a car that has gone over the jetty, we'll say, or a helicopter that has ditched or even to somebody who's a good swimmer who finds himself in the water adjacent to a pier or a jetty where the waves are rebounding off the jetty, he's getting rebound splash coming across his face, he won't be able to hold his breath and he'll aspirate water and he'll drown; so that's the first effect.

However, provided drowning did not occur, the next stage began:

After about two minutes – a minute to two minutes – that painful effect dies down and you can control your breathing again and now you just start feeling cold and you start shivering ... you were probably shivering earlier on but you're becoming conscious of it now because earlier on, you were thinking of your breathing and your heart rate. Now you're settling down, you're becoming aware of your situation, you're wondering how the hell you're going to get out of this and you're shivering like mad, and that shivering will build up in intensity until you're shivering to a level which is about half your maximum exercise capability which is quite a lot of oxygen to be using. And even in spite of producing all that heat inside your body, you're still not able to control the body temperature leaking out into this huge heat sink in which you're immersed and your body temperature will start falling.

At the same time as all of this is happening there is the pressure or the squeeze of the water around your body. If you can imagine putting your hand in a loosely fitting polythene bag, putting that hand into the water, you'll find that the polythene is squeezed to the skin around the hand, the air is squeezed out of the bag. That's the hydrostatic pressure around your body. Well, that's squeezing on the body tissues and it's squeezing the blood back into the heart. So if you can imagine now your heart as being a pump, there's more fluid coming in one side of the pump, it's going to go out the other side, so the output of your heart increases and you overload your circulation to a degree and the body will compensate for that by trying to get rid of some of that circulating fluid. So in fact you'll

> find you will want to urinate, you will want to offload some of this fluid and it's distributed in some of the other tissues centrally.

This, of course, has the effect of lowering body temperature by allowing warm fluids to escape into the cold surrounding water:

> Then as time goes on your shivering intensity is increasing, you're becoming semi-conscious. Eventually you will lose consciousness when you've lost about four or five degrees Centigrade from your body temperature and then you will just continue to drift off into hypothermia, deeper hypothermia, and you will eventually die when you lose about ten degrees body temperature, provided you can keep your airway out of the water. If you can't keep your airway clear of the water, well, then clearly you will drown much earlier and that will occur probably when you're just beginning to lose consciousness, when you're getting semi-conscious and that is when you've lost just about two or three degrees body temperature.

One of the survivors of the freezing experiments was Father Leo Miechalowski, a Catholic priest interned in Dachau as a political prisoner, who described his experience as one of Dr Rascher's experimental cases at the Nuremberg Trials after the war:

> On the seventh of October, 1942, a prisoner came and told me that I was to report to the hospital immediately. I thought I was going to be examined once more, and I was taken through the malaria station to block 5 in Dachau, to the fourth floor of block 5. There, the so-called aviation room, the aviation experimental station, was located there, and there was a fence, a wooden fence so that nobody could see what was inside, and I was led there, and there was a basin with water and ice which floated on the water. There were two tables, and there were two apparatus on there.

Next to them there was a heap of clothing that consisted of uniforms, and Dr Prachtol was there, two officers in Air Force uniforms. However, I do not know their names. Now I was told to undress. I undressed and I was examined. The physician then remarked that everything was in order. Now wires had been taped to my back, also in the lower rectum. Afterwards I had to wear my shirt, my drawers, but then afterwards I had to wear one of the uniforms which were lying there. Then I had also to wear a long pair of boots with cat's fur and one aviator's combination. And afterwards a tube was put around my neck and was filled with air. And afterwards the wires which had been connected with me – they were connected to the apparatus, and then I was thrown into the water. All of a sudden I became very cold, and I began to tremble. I immediately turned to those two men and asked them to pull me out of the water because I would be unable to stand it much longer. However, they told me laughingly, 'Well, this will only last a very short time.' I sat in this water, and I had – and I was conscious for one hour and a half. I do not know exactly because I did not have a watch, but that is the approximate time I spent there.

During this time the temperature was lowered very slowly in the beginning and afterwards more rapidly. When I was thrown into the water my temperature was lowered very slowly in the beginning and afterwards more rapidly. When I was thrown into the water my temperature was 37.6, then the temperature became lower. Then I only had 33 and then as low as 30, but then I already became somewhat unconscious and every fifteen minutes some blood was taken from my ear. After having sat in the water for about half an hour, I was offered a cigarette, which, however, I did not want to smoke. However, one of those men approached me and gave me the cigarette, and the nurse who stood near the basin continued to put this cigarette into my mouth and pulled it out again. I managed to smoke about half of this cigarette. Later on I was given a little glass with Schnapps, and then I was asked how I was feeling. Somewhat later still I was given one cup of Grog. This Grog was not very hot. It was rather lukewarm. I was freezing very much in this water. Now my feet were becoming as rigid as iron, and the same thing applied to my hands, and later on my breathing became very short. I once again began to tremble, and afterwards cold sweat appeared on my forehead. I felt as if I was just about to die, and then I was still asking them to pull me out because I could not stand this much longer.

Then Dr Prachtol came and he had a little bottle, and he gave me a few drops of some liquid out of this bottle, and I did not know anything about this liquid. It had a somewhat sweetish taste. Then I lost my consciousness. I do not know how much longer I remained in the water because I was unconscious. When I again regained consciousness, it was approximately between 8 and 8:30 in the evening. I was lying on a stretcher covered with blankets, and above me there was some kind of an appliance with lamps which were warming me.

In the room there was only Dr Prachtol and two prisoners. Then Dr Prachtol asked me how I was feeling. Then I replied, 'First of all, I feel very exhausted, and furthermore I am also very hungry.' Dr Prachtol had immediately ordered that I was to be given better food and that I was also to lie in bed. One prisoner raised me on the stretcher and he took me under his arm and he led me through the corridor to his room. During this time he spoke to me, and he told me, 'Well you do not know what you have even suffered.' And in the room the prisoner gave me half a bottle of milk, one piece of bread and some potatoes, but that came from his own rations. Later on he took me to the malaria station, block 3, and there I was put to bed, and the very same evening a Polish prisoner – it was a physician; his first name was Dr Adam, but I do not remember his other name – he came on official orders. He told me, 'Everything that has happened to you is a military secret. You are not to discuss it with anybody. If you fail to do so, you know what the consequences will be for you. You are intelligent enough to know that.' Of course, I fully realized that I had to keep quiet about that.

The freezing experiments continued until the spring of 1943, when they were broken off. The suggestion is that the sadism and brutality of Rascher's work was beginning to cause concern among the aviation medical experts who, along with Himmler, were sponsoring his research:

> Now when they got around to working out the different techniques for re-warming people who had frozen, who were hypothermic, one of the techniques that they used that had been theorized about in textbooks was using other human bodies: surrounding the individual with other human bodies so that in a combat situation your buddies could maybe surround you and warm you with their bodies and the question was whether that would work, if you didn't have access to electric blankets and so forth. So they tried that but Rascher tried it in a most sadistic and perverted way: he surrounded one of these subjects, a man, with nude women who were themselves concentration camp inmates and indeed the victim regained consciousness and began to copulate with one of the women and this was filmed and presented to Rascher's superiors whose puritanical – oddly puritanical – Nazi instincts revolted at this.
>
> And I think it was at that point that people began to change their minds decisively about Rascher: this violated their sensibilities – oddly in the context of death that was happening all around them – but this incident I think turned them off to Rascher and raised a question about his judgement because certainly that was not a necessary experiment.

Rascher had, however, sufficiently impressed Himmler for the Reichsführer-SS to use his influence to gain him an academic post at the University of Strassburg, but, in a startlingly bizarre ending to his story, he was arrested in 1944 for abducting children. In fact it emerged that Karoline Rascher was unable to have children and, in the interests of presenting themselves as the perfect Aryan family, Rascher had stolen two baby boys whom they attempted to pass off as their own. For this peculiar crime, Rascher was executed by the SS.

The brutality and sadism of Rascher's human experiments are very evident, and it has become customary since the end of war to dismiss them as having had little or no scientific value. Uncomfortably, this is not entirely true, as Frank Golden discovered when researching the same subject:

> Two things: firstly they were undoubtedly unethical, and it's a truism to say if it's unethical it's bad science. Now skipping that and looking at what we know of the results, whether the work was scientific or not is difficult, apart from the ethical issue. It's difficult to say because we haven't seen the scientific paper; all that is available to us is an executive summary that was produced by Major Alexander for the Nuremberg Trials, so we're just looking at the conclusions basically.
>
> Now, as to their accuracy, they did in fact throw a lot of light on this problem of hypothermia which wasn't known before, they gave a very good account of the symptomology that one encountered as the body cooled; they gave us a very good idea of what the lethal body temperature is, for example, which we weren't sure of before or people weren't sure of before; and they told us why the body, why you died in hypothermia. Prior to that there was an argument as to whether you died from a shortage of oxygen or whether it was the heart stopped. Those experiments shows us it was the stoppage or the cardiac arrest that was causing the problem. They revealed that in progressive cooling in this sort of situation the blood thickened quite a lot, they also showed that in this situation the blood sugar level rose, so there were a number of issues like that which we now know subsequently, in the light of post-war experiments – I hasten to say ethically conducted – and from

> information obtained from casualties, exposure casualties. We know that the results ... were largely correct, so they perhaps were scientifically obtained in that sense, although not ethically obtained, so that's quite a controversial issue as to whether you should use such information or not, I think.

In some ways this is the key to the matter. It has become commonplace to dismiss Rascher's experiments as being scientifically useless, but this is evidently not the case:

> Well, I've obviously thought long and hard about this... if it's unethical science you're not supposed to use the information. I think in this particular case that there were a number of people who unfortunately died to provide this information. We could sweep it under the carpet and forget all about it, but by using it at least I think it's acting as a memorial to those people. We're remembering them every time we use it, we're remembering the horror they had to go through, the unethical attitude of some of the scientists, if not all of them involved, and I think it would be morally wrong just to bury it and not use it. I think we acknowledge the work's done, it's there. I think perhaps we shouldn't acknowledge the name of the people involved but the fact that they did it, I think, yes, should be remembered.

Rascher was by no means the only doctor who experimented on human beings. Even while the low-pressure and freezing experiments were taking place at Dachau, some 1,200 inmates of the same camp were being infected with malaria by a Dr Schilling, who was experimenting with various combinations of anti-malarial drugs. One of the subjects of these experiments was Father Miechalowski, who also figured in Rascher's freezing experiments:

I was that weak that I fell down on the road because everybody was hungry in the camp. I wanted to be transferred to another assignment later on where we got some bread to eat between meals so my health could improve by the additional food. One man arrived and selected about thirty people for some easy labour. I also wanted to be selected for this assignment and those who had been selected for this work were led away. We went in the direction where the work was located and at the very last moment instead of going to the place of work we were lead to the camp hospital. We did not know what was going to be done with us there. I thought to myself that perhaps this was going to be some detail for easier work in the hospital. We were told that we should undress and after we had undressed ourselves our numbers were taken down and then we asked what was going on and they told us, smilingly, 'this is for a work detail.' But we were not told what was going to be done with us. Then the doctor came and told us all to remain and that we were to be X-rayed. Now that our numbers had already been taken down we were supposed to go to our blocks. I sat for two days in the block and afterwards I was again called to the hospital and there I was given malaria in such a manner that there were little cages with infected mosquitoes and I had to put my hand on one of the little cages and a mosquito stung me and afterwards I was still in the hospital for five weeks. However, for the time being no symptoms of the disease showed themselves. Somewhat later, I don't exactly recall, two or three weeks, I had my first malaria attack. Such attacks recurred frequently and several medicines were given to us against malaria. I was given such medicine as Neosalvarsan. I was given two injections of quinine. On one occasion I was given antipyrin and the worst was that one time when I had an attack, I was given so-called pyramidon. I was given nine injections of that kind, one every hour and that every second day through the seventh injection. All of a sudden my heart felt like it was going to be torn out. I became insane. I completely lost my language – my ability to speak. This lasted until evening. In the evening a nurse arrived and wanted to give me the eighth injection. I was then unable to speak and I told the nurse about all of the complications I had had and that I did not want to receive the injection. The nurse had already poured out the injection and said that he would report this to Dr Schilling. After approximately ten minutes another nurse arrived and said that he would have to give me the injection after all. Then I said the same thing again, that I was not going to have the injection. However, he told me that he had to carry out that order. Then I replied that no matter what order he had, I would not be willing to commit suicide. Then he went away and returned once again after ten minutes. He told me, 'I know you know what can happen if you don't accept the injection.' Then I said in spite of everything, 'I refuse to receive a another injection and that I would tell that to the professor.' I requested that he himself know that I would not be willing to receive the injection. So that the nurse would not have any further difficulty, after twenty minutes Dr Ploettner came with four inmate nurses and he talked to my comrades. 'There is going to be a big row here.' Then I said, 'If I have resisted for such a long time I will continue to do so.' Dr Ploettner, however, was very quiet. He only reached for my hand and he checked my pulse, then touched my head and asked me what complications I had had. I told him what I had had after that injection. And then he told the nurse to give me two tablets in order to remove the headache and the pains in my kidneys. When I had been given that Dr Ploettner was about to leave and told the nurses that they were to give me the rest of the injections. Then I said, 'Hauptsturmführer, I refuse to be given that injection.' The physician turned around after I had said that and looked at me and said, 'I am responsible for your life, not you.' 'Then give the injection,' he told the nurse – the nurses complied with his order and it was then they gave me this injection. It was the same one to whom I had previously told that I did not want to have another injection. It

> was only strange that after the eighth injection no results happened as they had done previously so that, in my opinion, I think that the nurse gave me some other injection. On the morning I was given the ninth injection – when I woke up in the morning the results were then as usual. I became sick and I began to feel cold and I had a high fever.

Possibly the most barbaric of all the experiments were perpetrated on inmates from the women's concentration camp at Ravensbrück in northern Germany. These were designed to study bone, muscle and nerve regeneration after battlefield injury, and the effects of sulphanilamide drugs. The victims of these experiments were wounded and then deliberately infected with a range of bacteria, and then treated with drugs to gauge their impact. Naturally, in a scientific experiment, there was a control group who were not given the drugs and, equally naturally, many of this group died. One of the victims, Vladislawa Karolewska, a Polish woman detained because she was suspected of assisting partisans, described what had happened to her to the court at Nuremberg:

> On the 14th of August, the same year, I was called to the hospital and my name was written on a piece of paper. I did not know why. Besides me, eight other girls were called to the hospital. We were called at a time when usually executions took place and I thought I was going to be executed because before some girls were shot down. In the hospital we were put to bed and the hospital room in which we stayed was locked. We were not told what we were to do in the hospital and when one of my comrades put the question she got no answer but she was answered by an ironic smile. Then a German nurse arrived and gave me an injection in my leg. After this injection I vomited and I was put on a hospital cot and

> they brought me to the operating room. There, Dr Schildauski and Rosenthal gave me the second intravenous injection in my arm. A while before, I noticed Dr Fischer who went out of the operating room and had operating gloves on. Then I lost my consciousness and when I revived I noticed that I was in a regular hospital room. I recovered my consciousness for a while and I felt severe pain in my leg. Then I lost my consciousness again. I regained my consciousness in the morning and then I noticed that my leg was in a cast from the ankle up to the knee and I felt a very strong pain in this leg and the high temperature. I noticed also that my leg was swollen from the toes up to the groin. The pain was increasing and the temperature, too, and the next day I noticed that some liquid was flowing from my leg. The third day I was put on a hospital cart and taken to the dressing room. Then I saw Dr Fischer again. He had an operating gown and rubber gloves on his hands. A blanket was put over my eyes and I did not know what was done with my leg but I felt great pain and I had the impression that something must have been cut out of my leg.

In fact, using anaesthetic, the doctors were attempting to simulate a battle wound:

> After the changing of the dressing I was put again in the regular hospital room. Three days later I was again taken to the dressing room, and the dressing was changed by Dr Fischer with the assistance of the same doctor, and I was blindfolded, too. I was then sent back to the regular hospital room. The next dressings were made by the camp doctors. Two weeks later we were all taken again to the operating room and put on the operating tables. The bandage was removed, and that was the first time I saw my leg. The incision went so deep that I could see the bone. We were told ... a doctor from Hohenlychen, Doctor Gebhardt, would come and examine us. We were

waiting for his arrival for three hours lying on our tables. When he came a sheet was put over our eyes, but they removed the sheet and I saw him for a short moment. Then, we were taken again to our regular rooms. On the eighth of September I was sent back to the block. I could not walk. The pus was draining from my leg; the leg was swollen up and I could not walk.

In the block, I stayed in bed for one week; then I was called to the hospital again. I could not walk and I was carried by my comrades. In the hospital I met some of my comrades who were there for the operation. This time I was sure I was going to be executed because I saw an ambulance standing before the office which was used by the Germans to transport people intended for execution. Then, we were taken to the dressing room where Doctor Oberhauser and Doctor Schildauski examined our legs. We were put to bed again, and on the same day, in the afternoon, I was taken to the operating room and the second operation was performed on my leg. I was put to sleep in the same way as before, having received an injection. And, this time I saw again Doctor Fischer. I woke up in the regular hospital room and I felt a stronger pain and higher temperature.

The symptoms were the same. The leg was swollen and the pus flowed from my leg. After this operation, the dressings were changed by Dr Fischer every three days. More than ten days afterwards we were taken again to the operating room, put on the table; and we were told that Dr Gebhardt was going to come to examine our legs. We waited for a long time. Then he arrived and examined our legs while we were blindfolded. This time other people arrived with Dr Gebhardt; but I don't know their names; and I don't remember their faces. Then we were carried on hospital cots back to our rooms. After this operation I felt still worse; and I could not move. While I was in the hospital, cruelty from Dr Oberhauser was performed on me.

They were not only being mutilated by the doctors, but also punished when they complained about it:

> When I was in my room I made the remark to fellow prisoners that we were operated on in very bad conditions and left here in this room and that we were not given even the possibility to recover. This remark must have been heard by a German nurse who was sitting in the corridor because the door of our room leading to the corridor was opened. The German nurse entered the room and told us to get up and dress. We answered that we could not follow her order because we had great pains in our legs and we couldn't walk. Then the German nurse came with Dr Oberhauser into our room. Dr Oberhauser told us to dress and come to the dressing room. We put on our dresses; and, being unable to walk, we had to hop on one leg going into the operating room. After one hop, we had to rest. Dr Oberhauser did not allow anybody to help us. When we arrived at the operating room, quite exhausted, Dr Oberhauser appeared and told us to go back because the change of dressing would not take place that day. I could not walk, but somebody, a prisoner whose name I don't remember, helped me to come back to the room.

Despite the likely repercussions, the victims of these experiments were sufficiently outraged to complain officially about their treatment:

> At the end of February 1943, Dr Oberhauser called us and said, 'Those girls are now guinea-pigs', and we were very well known under this name in the camp. Then we understood that we were persons intended for experiments and we decided to protest against the performance of those operations on healthy people.
>
> We drew up a protest in writing and we went to the camp commander.

Not only those girls who had been operated on before but other girls who were called to the hospital came to the office. The operated-on girls used crutches and they went without any help.

I would like to tell the contents of the petition made by us. We, the undersigned, Polish political prisoners, ask Herr Commander whether he knew that since the year 1942 in the camp hospital experimental operations have taken place under the name of guinea-pig (*das sind Meerschweine*), as explaining the meaning of those operations. We ask whether we were operated on as a result of sentences passed on us because, as far as we know, the international law forbids the performance of operations even on political prisoners.

Predictably:

We did not get any answer; and we were not allowed to talk to the commander. On the 15th of August, 1943, a police woman came and read off the names of the ten new prisoners. She told us to follow her to the hospital. We refused to go to the hospital, as we thought that we were intended for a new operation. The police woman told us that we were going probably to be sent to a factory for work outside the camp. We wanted to make sure whether the Arbeitsamt (work centre) was open because it was Sunday. The police woman told us that we had to go to the hospital and be examined by a doctor before we went to the factory. We refused to go then because we were sure that we will be kept in the hospital and operated on again. All prisoners in the camp were told to stay in the blocks. All of the women who lived in the same block where I was were told to leave the block and stand in line before the block ten at a time. Then overseer Binz appeared and called out ten names and among them was my name. We went out of the line and stood before the ninth block in line. Then Binz said: 'Why do you stand so in line as if you were to be executed?' We told her that the operations were worse for us than executions and that we would prefer to be executed rather than to be operated on again. Binz told us that she might give us work, there was no question of our being operated on but we were going to be sent for work outside the Camp. We told her that we must know that prisoners belonging to our group are not allowed to leave the camp and go outside the camp. Then she told us to follow her into her office, that she would show us a paper proving that we are going to be sent for work to the factory outside the camp. We followed her and we stood before her office. She entered her office for a while and then went to the canteen where the Camp Commander was. She had a conference with him probably asking him what to do with us. We stood before the office a half an hour. In the meantime one fellow-prisoner who used to work in the canteen walked by us. She told us that Binz asked for help from SS men to take us by force to the hospital. We stood for a while and then Binz came out of the canteen accompanied by the Camp Commander. We stood for a while near the camp gate. We were afraid that SS men would come to take us so we ran away and mixed with other people standing before the block. Then Binz and the camp police appeared. They drove us out from the lines by force. She told us that she put us into the bunker as punishment; that we did not follow her orders. In each cell were put five prisoners although one cell was intended only for one person. The cells were quite dark; without lights. We stayed in the bunker the whole night long and the next day. We slept on the floor because there was only one couch in the cell. The next day we were given a breakfast consisting of black coffee and a piece of dark bread. Then we were locked again in this dark room. We were only troubled by people walking in the corridor of the bunker. The answer was given us the same day in the afternoon. The watch-woman from the bunker unlocked our cell and got me out of the cell. I thought

> that I was then to be interrogated or beaten. They took me and they went down the corridor. She opened one door and behind the door stood SS man Dr Trommel. He told me to follow him upstairs. Following Dr Trommel I noticed there were other cells, and those cells were with bed clothing. He put me in one of the cells. Then he asked me whether I would agree to a small operation. I told him that I did not agree to it because I had undergone already two operations. He told me that this was going to be a very small operation and that it will not harm me. I told him that I was a political prisoner and that the operation cannot be performed on political prisoners without their consent.

In fact the doctors preferred to experiment on political prisoners as they were generally more healthy than, for example, Jews, who had been on starvation rations for much longer:

> He told me to lie down on the bed; I refused to so. He repeated it twice. Then he went out of the cell and I followed him. He went quickly downstairs and locked the door. Standing before the cell I noticed a cell on the opposite side of the staircase, and I also noticed some men in operating gowns. There was also one German nurse ready to give an injection. Near the staircase stood a stretcher. That made it clear to me that I was going to be operated on again in the bunker. I decided to defend myself to the last moment. In a moment Trommel came with two SS men. One of these SS men told me to enter the cell. I refused to do it, so he forced me into the cell and threw me on the bed.
>
> Dr Trommel took me by the left wrist and pulled my arm back. With his other hand he tried to gag me, putting a piece of rag into my mouth, because I shouted. The second SS man took my right hand and stretched it. Two other SS men held me by my feet. Immobilized, I felt

> that somebody was giving me an injection. I defended myself for a long time, but then I grew weaker. The injection had its effect; I felt sleepy. I heard Trommel saying, '*Das ist fertig*' ['It's ready'], that is all.
>
> I regained consciousness again, but I don't know when. Then I noticed that a German nurse was taking off my dress, I then lost consciousness again; I regained it in the morning. Then I noticed that both my legs were in iron splints and were bandaged from the toes to groin. I felt a strong pain in my feet, and a temperature.

Just as with the euthanasia programme, the horrifying fact remains that much of the brutality was carried out not by hardened SS men, but by doctors and nurses who had previously devoted all of their professional attention to curing their patients. Even when the doctors did happen to be members of the SS, like Rascher (although in a strictly part-time manner), it would not be unreasonable to assume that their professional training would override political demands to injure and kill people. Evidently this was not the case, and there is much to suggest that many of the research projects which used human subjects were generated by the doctors themselves.

The most notorious of the Nazi medical experimenters, Dr Josef Mengele, the son of a wealthy farm-machine manufacturer from Günzburg in southern Germany, had served on the eastern front as a Waffen-SS medical officer, before being posted to the concentration camp at Auschwitz as part of the medical staff. Apart from his role in making selections of inmates when they arrived, deciding who would live and who would die, he took it upon himself to harvest organs from twins and dwarfs who passed through his hands to send as experimental samples to his academic mentor, Professor Ottmar Verschuer, at the Kaiser

Wilhelm Institute for Hereditary Biology in Berlin. This was not done out of sadism, as has often been claimed, but because Mengele hoped to ingratiate himself with Verschuer in order to secure a prestigious academic post.

Even for doctors, National Socialism successfully turned people into a commodity.

CHAPTER FIVE

nuclear fission:

the jewish science

On 6 August 1945, at Farm Hall, a large country house just outside Cambridge, a strained conversation was taking place between a group of distinguished scientists.

'...If the Americans have a uranium bomb then you're all second-raters. Poor old Heisenberg,' announced Otto Hahn, a Nobel Prize-winning radiochemist and the discoverer of nuclear fission.

'Did they use the word uranium in connection with this atomic bomb?' asked Werner Heisenberg, another Nobel Prize-winner and the discoverer of quantum mechanics. On being assured that this was not the case, he continued: '...Then it's got nothing to do with atoms, but the equivalent of 20,000 tons of high explosive is terrific... All I can suggest is that some dilettante in America who knows very little about it has bluffed them in saying, "If you drop that it has the equivalent of 20,000 tons of high explosive" and in reality it doesn't work at all.'

Hahn was unconvinced: 'At any rate, Heisenberg, you're all just second-raters and you may as well pack up.'

'I quite agree,' added Heisenberg ruefully.

The event they were discussing was the American atomic bombing of Hiroshima which had taken place that morning. It was not yet clear what the result had been, as smoke and cloud still obscured the Japanese city, but it was evident that a new era in warfare had begun. The reason for the general air of incredulity with which the news had been greeted was that the group of men were Germany's leading nuclear scientists, most of whom had been involved in research aimed at building an atomic bomb for the Third Reich. Interned in England and kept under close surveillance at all times, they were being held 'on ice' while British and US scientific and intelligence specialists assessed the German bomb project and whether any of the scientists was likely to be useful to the Soviets if they were allowed to go free.

In fact detailed investigation by an Allied special scientific mission had already revealed the basic truth of German atomic-bomb research: it had failed, despite the presence of Heisenberg, universally regarded as one of the foremost theoretical physicists of the age and a man who would, under different circumstances, have certainly been recruited to work on the Manhattan Project, the huge operation which had built the atomic bomb for the USA. Germany's scientists hadn't even managed to get a nuclear reactor to work, and they found it hard to believe that the Americans and British had succeeded. The reasons for their failure were mainly technical, but underlying these was the catastrophic impact that the Third Reich had on the German scientific community.

The realization that it might be possible to create a bomb which would utilize the extraordinary energy of a fissioning atom had burst on to the scientific world in the spring of 1939, following the discovery of nuclear fission itself the previous Christmas. Nevertheless, the first forty years of the twentieth century had seen a series of extraordinary and revolutionary advances in science's understanding of the structure of the atom and its component particles which had made this inevitable.

Ideas that matter might be composed of tiny fundamental particles – atoms – go back to ancient Greece. In the fifth century BC, Leucippus of Miletus and his follower Democritus hypothesized that such properties as colour, hardness, mass and heat were the result of different arrangements of tiny fundamental particles which they described as *atomos*, or 'indivisible'. These ideas were picked up and toyed with by a number of subsequent classical philosophers, but the lack of experimental data made them impossible to prove, even though they provided a reasonable explanation for many commonly observed phenomena, and so they languished as an intellectual dead end until the rise of empirical science towards the end of the eighteenth century. Until then science had been simply the application of logic to the phenomena of everyday life, but the new generation of scientists who emerged at this time – Lavoisier, Black, Priestley, Klaproth, Cavendish and so forth – applied critical logic to empirical experimentation and began to make major inroads into understanding the physical composition of the world about us. Atomic theory was revived and systematized by James Dalton in the early nineteenth century and by mid-century it had been accepted as a fundamental underpinning of the physical universe.

The latter part of the nineteenth century saw research into the structure and characteristics of the atom proceed at some speed. In 1895 the German physicist Wilhelm Konrad Röntgen, while studying cathode rays in his laboratory at Würzburg, had – by chance – observed a strange luminescence on chemically coated papers he was using in his experiments. Even more surprising was the observation that the same effect was visible in an adjacent room. Röntgen named the agent for this effect 'X-rays' and was soon using them to take 'shadow photographs' of apparently solid structures, including the human body. This discovery was followed shortly afterwards by Max Planck's quantum theory, by the discovery of subatomic particles and the 'nuclear' theory of the atom, and by Albert Einstein's special and general theories of relativity (in 1905 and 1916 respectively), which included, in the special theory, the famous equation for the conversion of matter into energy: $E=mc^2$ where 'E' is energy, 'm' is mass and 'c' is the speed of light multiplied by itself. This implies that if a relatively small mass is destroyed by conversion into energy, a very large amount of energy will be released.

Other research at this time began to indicate that the atomic structures of some of the heaviest atoms were also the most unstable. The French physicist Antoine Henri Becquerel was conducting experiments into X-rays when he discovered that uranium salts would darken photographic plates as if they had been exposed to light, and that this effect appeared to be independent of external factors. This research was taken up by Marie Curie, a Polish-born physicist working in Paris, who exhaustively tested other materials for radioactivity. The only material she found was pitchblende, the ore from which uranium is extracted, but this appeared to be even more radioactive than pure uranium. Hypothesizing that another radioactive element must be present, Marie Curie and her husband, Pierre, methodically refined and separated the pitchblende until eventually they had discovered not one new radioactive element, but two: radium and polonium. One of the most interesting aspects of these new elements was the large amount of energy available from radioactivity: certainly considerably more than would be produced by a chemical reaction in the same weight of material.

Related research by the New Zealander Ernest Rutherford in the first decade of the twentieth century further elucidated the structure of the atom, demonstrating that it consisted, in part at least, of a tiny but massively dense nucleus carrying positive electrical charges, around which, at some distance, spiral electrons with a negative charge equal to the positive charge at the centre. This was a good explanation of the structure of the atom but not a complete one, which would have to wait until the First World War had ended.

Rapid progress resumed after the First World War with the important discovery, by the Englishman James Chadwick (who had been interned in Germany throughout the war), of the neutron, a particle which had the same mass as a proton but no electrical charge and which would, therefore, be able to penetrate the nucleus of an atom. In the early 1930s the Italian Enrico Fermi began experimenting with bombarding atomic nuclei with neutrons. When he tried this with uranium he was puzzled to find that the nuclei 'captured' the neutrons, thus apparently gaining one unit of atomic weight. If this was true, the new 'isotope' of uranium 239 would then shed a beta particle and transmute into element 93 – a substance that was not known to exist on earth. The products of Fermi's neutron bombardment did not seem to behave like other known heavy elements – the result that might be expected – and he became increasingly convinced that he was creating new 'transuranic' elements.

Fermi's apparent discovery – for which he was awarded the Nobel Prize – was naturally a cause of great excitement among physicists and chemists, and throughout the latter half of the 1930s they laboured to nail down what the products of neutron bombardment of uranium actually were. There was some speculation that, in addition to element 93, uranium was giving birth to elements 94, 95 and even 96, and even those who questioned the likelihood of transuranic materials being produced were at a loss to explain what was occurring. At one time it was hypothesized that element 91, protactinium, was being produced, but when this was tested by the radiochemist Otto Hahn, who had discovered it in the first place, this was found not to be the case.

One dissenting voice in the general climate of opinion on Fermi's work was raised by a Hungarian chemist working in

Germany in the mid-1930s. Ida Noddack suggested that by limiting their research to the rare earth and heavy elements, which they assumed would be the most likely products of uranium bombardment, scientists were leaving huge areas unexplored. She thought it possible that the nuclei of heavy atoms broke into large fragments which would be isotopes of completely unrelated elements. Although this view had logic on its side, the nuclear physics 'establishment' widely dismissed her as cranky and wrong.

Nevertheless, Noddack had hit the nail on the head. Researchers increasingly began to garner evidence that the products of uranium bombardment came from much further down the periodic table of elements, rather than at the uranium end. Irene Curie, daughter of Marie Curie, tentatively suggested that one such product was 'like lanthanum' (element 57) without ever realizing the truth (in fact it *was* lanthanum). Hahn, at the Kaiser Wilhelm Institute in Berlin, decided to follow up this suggestion and came to a startling conclusion which he felt violated all previous experimental evidence in nuclear physics; it appeared that, under neutron bombardment, uranium had transmuted into two much lighter elements, barium and lanthanum. This seemed almost alchemical: it should not be possible to turn one element into two completely different ones.

So great were the misgivings of Hahn and his colleague Fritz Strassman that their report was couched in the most neutral terms and, even then, Hahn felt the need to discuss his findings with his old and valued friend and colleague, Lise Meitner. To do this, Hahn, ludicrously, had to expose himself to a certain amount of risk: Meitner was an Austrian-born Jew and, despite the fact that she had lived and worked in Berlin since 1907, had been facing increasing persecution while she remained there. Until March 1938 her Austrian nationality had served to protect her from the worst excesses of Nazi-inspired anti-Semitism, but following the Anschluss – Germany's annexation of Austria – she had suddenly become a German citizen (insofar as Jews could acquire German citizenship). The result of this was that she had to wear the yellow star and soon was subjected to verbal and physical assaults in public.

In some respects Meitner was fortunate. Her reputation as a scientist was such that she would have little difficulty finding a job outside Germany, but for some months she clung to the hope that she would be able to remain in the country. While she did so, what has been described as an 'international scientific underground' of anti-Nazi academics around Europe conspired to rescue her. In July 1938 she departed, aided by the promise of a trouble-free crossing into the Netherlands and a job in Stockholm; she had left Germany just in time.

So, when Hahn wrote to her just before Christmas 1938, he was risking having his letter intercepted by the state authorities for communicating with a Jew and presumptive enemy of the state. The problem he outlined was this: '...there's something so odd about the "radium isotopes" that for the moment we don't want to tell anyone but you... Our radium isotopes behave like barium... Perhaps you can put forward some fantastic explanation... We realize it [uranium] can't really burst into barium ... but we must clear this thing up.' He followed this up with a copy of a paper that he and Strassman had drafted for the scientific journal *Die Naturwissenschaften.*

For several years Meitner had been accustomed to spending her Christmas holiday with her nephew Otto Frisch, also a refugee from Nazism and then working with the Dane Niels Bohr, the presiding genius of nuclear physics, in Copenhagen. On Christmas Eve 1938 they went for a walk in the snow in the countryside surrounding the small Swedish seaside resort of Kungälv, where they were staying. Frisch was initially sceptical of Hahn's results but after some discussion the two stopped to sit on a fallen tree trunk and, using the back of Hahn's letter to make notes, they worked out their interpretation of his findings. Bohr had, in the previous few years, been refining the idea that the nucleus of an atom was similar to a droplet of liquid. Applying this idea to Hahn's odd results, Frisch suddenly began to see how the process must work:

> A nucleus was not like a brittle solid that can be cleaved or broken ... a nucleus was much more like a liquid drop. Perhaps a drop could divide itself into two smaller drops in a more gradual manner, by first becoming elongated, then constricted, and finally being torn in two? We knew that there were strong forces that would resist such a process, just as the surface tension of an ordinary liquid drop tends to resist its division into two smaller ones. But the nuclei differed from ordinary drops in one important way: they were electrically charged, and that was known to counteract surface tension.
>
> ...so the uranium nucleus might indeed resemble a very wobbly, unstable drop, ready to divide itself at the slightest provocation, such as the impact of a single neutron.

Frisch and Meitner also worked out that, as they separated, the two drops would acquire a very high speed and thus a large energy. Where would this have come from? Meitner calculated that about one fifth of the mass of a proton would be lost during the division and, using Einstein's formula for the conversion of mass into energy, $E=mc^2$, they found that it seemed to fit. In fact this 'fission' of one uranium atom would release sufficient energy to visibly move a grain of sand. Excited by their discovery, Frisch returned to Copenhagen to tell Bohr.

It so happened that Bohr was about to depart for the USA, where he was due to give a series of lectures and addresses to nuclear physicists, and anxious to ensure that Frisch and Meitner were given credit for their interpretation of Hahn's experiments, he duly announced what they had discovered. It was a discovery of enormous moment.

The idea of using the energy of fissioning atoms had been around for some years – HG Wells had written a book about it in 1909 – and various uses had been contemplated, ranging from engines to bombs. The realization that nuclear fission had been achieved threw it all suddenly into sharp focus. A Hungarian Jewish émigré, Leo Szilard, who some years before had conceived the idea of a nuclear chain reaction, patented it and assigned the patent to the British Admiralty, wrote to a friend that the discovery might: '...lead to large-scale production of energy and radioactive elements, unfortunately also perhaps to atomic bombs. This new discovery revives all the hopes and fears in this respect which I had in 1934 and 1935, and which I have as good as abandoned in the last two years.'

Together with Enrico Fermi (who, after being awarded the Nobel Prize, had left Italy because his wife was Jewish) and Walter Zinn, a colleague of Fermi, Szilard hired the necessary equipment to bombard natural uranium with neutrons. Szilard and Fermi wanted to see how many neutrons would be emitted when uranium was itself bombarded by them; these neutrons would show up to them as flashes on a television screen. Szilard later wrote: 'We turned the switch and we saw the flashes. We watched for a while and then we turned off the machine and went home. That night, there was little doubt in my mind the world was headed for grief.'

The concept is quite simple. If an atomic nucleus fissions and releases one or two neutrons and you have a big enough piece of uranium, it is likely that one of these neutrons will hit and fission another nucleus, and so on, causing a continuous chain reaction that will multiply, giving out increasingly large amounts of energy. They had proved, to their own satisfaction, that a self-sustaining chain reaction might be caused which would liberate a staggering amount of energy. Other scientists were thinking on the same lines. In California the news from the east coast caused a considerable stir. Philip Morrison was doing research under Robert Oppenheimer at the University of California at Berkeley, in San Francisco:

> I was a graduate student, maybe in my third or fourth year at Berkeley, so I was a grown-up graduate student and we'd all been puzzled in the previous weeks by the famous paper by Otto Hahn and Strassman – very good radiochemists – who discovered this phenomenon that when they irradiated uranium they got some ... element like krypton or something like that. We couldn't understand it, nobody could understand it, a couple of people gave talks about it but they couldn't understand it.
>
> There was a meeting in the east and a few people from Berkeley had gone there, not me, no students, but just a few of the professors went and during ... the last week of January or the very first week of February 1939 ... somebody called up and news spread right around the lab. Bohr had come to this meeting – we knew he was going to be there – and he brought the paper of Meitner and Frisch and told about it and it was a

sensation and they saw it demonstrated, right away.

...two or three people had neutron sources and scopes and chambers and so on, [and] went to work and I think that day or maybe the next day they were showing it off – it took a couple or three days to get around to me ... but pretty soon somebody said 'come up and see it, it's exciting', so I went there and I saw this amazing thing. Here was the oscilloscope screen and there were all these little green pulses on it and those little green pulses were started up by the power of the alpha particles of uranium, so they were at the bottom of the screen like blades of grass growing and then suddenly a great big streak would shoot up, much denser and much higher than any of the others. That was the fission ... it had 100 times more energy than the others, that's all you could say.

...we went back to the office ... we had a big office with six or eight desks in it right next to Robert Oppenheimer's small office, which was a private office for himself with maybe another desk for somebody. Well, in the student office we talked, of course, quite a lot and by the end of the next day or two we had drawn on the blackboard, believe it or not, what I've often said was a caricature of an atomic bomb.

Well, we knew very little about it but we understood the fact that there was fission, the nucleus was tremendously disturbed, it gave a lot of energy, it probably gave out neutrons ... we didn't know that – we knew very little about neutrons but it was a good bet, if it gave out more than one neutron, it could start a chain reaction, that was obvious, in a big enough piece, so you would say that, well, this might lead to a macroscopic size, that is a pint-sized or a pound-sized nuclear reaction which would be a devastating explosion ... so we drew this thing with heavy water and uranium and so on. We didn't understand it very well but, as I say, it was a caricature: you couldn't make one from that but you were in the right league ... it wasn't making violins or transistors, it was certainly making bombs... I learned only after the war when I read the

> publication of Robert Oppenheimer's letters that he had written a letter very much like that ... about the same time, to his old friend Uhlenbeck, in Michigan, telling him what he thought. He had this remarkable phrase in it: 'it looks very much as though a kilogram of uranium deuteride might blow itself to hell'.

Not surprisingly, speculation of this kind was taking place inside the Third Reich as well. Indeed it would have been far more surprising had it not, for Nazi Germany remained the home of one of the towering geniuses of twentieth-century nuclear physics, Werner Heisenberg.

Born in Munich in 1901, Heisenberg had been too young to serve in the First World War, but he had found himself afterwards serving in a Freikorps, a right-wing militia formed to help crush the communist revolution which had briefly seized control in his home city. Heisenberg studied physics at Munich under Professor Arnold Sommerfeld, but his scientific career blossomed in the early 1920s when he became part of the circle surrounding Niels Bohr in Copenhagen. In some respects Bohr was the bridge between the old Newtonian classical physics and the new quantum theory which was beginning to give insights into the behaviour of matter at the subatomic level. But Heisenberg was definitely part of the new physics.

Heisenberg secured his reputation with two specific contributions to physics: the first of these was his quantum mechanics. The problem facing physicists was that classical physics was failing to provide answers about the behaviour of particles at the subatomic level. Classical physics demanded universal laws and models to provide an explanation, but Heisenberg began to understand that these might be unnecessary; that in fact mathematical formulae alone could provide solutions, independent of intuitive analogies. In 1925 he developed these ideas into a system which became known as 'matrix mechanics'. This was not a simple solution, but for those with the mathematical skills to understand it, it did appear to point the way to a solution. At about this time a friend of Heisenberg, Wolfgang Pauli, had written in a letter to a friend: 'At the moment physics is again terribly confused. In any case, it is too difficult for me, and I wish I had been a movie comedian or something of the sort and had never heard of physics'. However, only five months later, he was able to write: 'Heisenberg's type of mechanics has again given me hope and joy in life. To be sure it does not supply the solution to the riddle, but I believe it is again possible to march forward.'

What quantum mechanics did was provide a model for the structure of the atom which worked as a series of immensely complex mathematical formulae. For those who could grasp it, it was far more useful than attempts to visualize the atom like a clockwork mechanism in three dimensions.

Heisenberg's second major contribution at this point was his 'Uncertainty Principle'. Part of the attack that was launched against his matrix mechanics was a theory that subatomic particles might also behave like waves. The Austrian physicist Erwin Schrödinger came up with a simple and elegant mathematical tool – wave mechanics – which appeared to offer a solution to many of the problems of classical physics as well as eliminating some of the more startling and inexplicable phenomena demanded by quantum theory. In fact Heisenberg recognized that Schrödinger's waves were really an expression of probability about the position of a given particle at a given time.

He saw that it is impossible, at any given instant, to discover both the speed and position of a particle, because to discover one, you have to change the other. The importance of probability in understanding the physical universe was not appreciated at the time: Einstein memorably objected: 'God does not play dice with the universe'; but equally Bohr – who supported Heisenberg and acted as his intellectual father figure – could counter: 'It is not our business to prescribe to God how he should run the world.'

The international community of nuclear physicists was small in the pre-war period: no more than a few hundred established scientists, who, together with their students, made up a network whose centre of gravity was, to a surprisingly large extent, in Germany. The reason for this lay in the structure of the German academic system, according to Professor Gerald Holton:

> ...they had very good structure. It was a division between various universities, unlike France, for example, where if you were not in Paris you were likely to be thought of as being in a second-rate place... The German universities were competing with each other, so Göttingen and Heidelberg and Berlin – and even Kiel and other places – were each trying very hard to get on top of things. And this competition, which was analogous to what was happening in America a bit later, really produced the atmosphere in which talent was badly needed... There was always competition between them [but] also collaboration: that is, there was a gentlemanly way of dealing with each other in large associations, in colloquia, in symposia. Von Laue, for example, ran a symposium in Berlin in which in the front row were all the great scientists from all over Germany at that time, and this combination of competition and collaboration produced a very spectacular, let us say, critical mass effect which was lacking in the United States.

The relative smallness of the nuclear physics 'club' was transcended by its importance. In 1927 the Solvay Conference of nuclear physicists took place in Brussels and several group photographs were taken. In a sense these pictures are a unique landmark in the history of science, as Holton explains:

> ...in that picture you may remember you have Heisenberg standing in the back row toward the right, hardly visible, very young. In the front row are the older physicists, Einstein and Franck and so on ... and it is a moment in which physics is turning and it is turning because of Born, because of Heisenberg, because of the Copenhagen group which is part of it. And on the other side are the older Germans such as Einstein himself and some of his friends, Schrödinger and so on, and they are still in the classical mode, that isn't to say that they are Newtonians but rather that they are trying to see whether the old physics can be tweaked to deal with the new effects, whereas in the background Heisenberg stands there as the true revolutionary and he is going to change everything...

Heisenberg's career prospered in Germany in the pre-Nazi era – not surprisingly, his quantum mechanics was one of the major tools that helped to push nuclear physics forward, and he was a well-respected teacher and academic leader – but there were non-scientific problems faced by all the new breed of theoretical physicists. In part this was caused by the sheer complexity of what they were doing. Many of the older classical experimental physicists found it very difficult to adapt to the new ideas and the new way of making progress, through blackboards covered in strange mathematical notations rather than experiments in the laboratory, and some of them became very bitter that physics seemed to have entirely passed them by.

One such was the Nobel Prize winner Philipp Lenard, who taught at Heidelberg. Lenard hadn't taken long to notice that Einstein and many of the other leading theoretical physicists were Jewish (or of Jewish extraction) and, having lost the scientific arguments, he began to attack the practice of 'Jewish physics'. In coining this meaningless concept, Lenard was simply lashing out at a branch of knowledge that he was unable to keep up with. Instead he chose to contrast 'Jewish physics' with 'German physics', which clung steadfastly to the intuitive 'natural' world. According to the historian Thomas Powers: '"Jewish physics" joined "Jewish art" and "Jewish literature" in the National Socialist lexicon of modernist threats to traditional German values. Initially discounted by Heisenberg and others as crackpot ravings, Nazi racism gradually forced its way into German universities.' This is actually a perfect example of the way that Nazism appealed to a wide range of embittered and dispossessed minorities: the original thrust of Nazism was the disbelief of a group of ill-educated, bellicose, reactionary beer-hall thugs that Germany could have been defeated in the First World War. This was translated into a search for 'external' enemies, like the Jews and the communists, whose machinations could explain this defeat.

Heisenberg was German and proud of it, but he was no Nazi, and he saw himself in German terms as part of the intellectual elite. Holton describes him thus:

> He was what one might call one of the best models of a 'good German'. He himself said: 'I embrace the Prussian virtues, which are modesty in living, being reliable, being punctual.' All the good things, and he really lived up to this idea. This is one aspect, the other is that he regarded himself – as the

> best Germans did – as a Kulturträger as he put it ... that is, a carrier of culture. He was not just interested in his narrow speciality, he also was widely read. He, with his seven children, had nightly concerts at home – music was extremely important to him, he was a very good pianist – and he regarded himself not as a mandarin in the German sense of serving the state but rather as a person that would further the mission of Germany to be a model for culture. And that was very appealing because there was a great deal in German culture before the Nazis took over that was worth promoting.

Moreover, Heisenberg was widely liked by his students and colleagues. Hans Bethe met him at Munich University, where Heisenberg was a regular visitor, when he was doing postgraduate research and working as a post-doctoral research fellow: 'Heisenberg was a very open person, very easy to talk to. An excellent physicist, of course, and with his students and collaborators, he was just a brother, so he was very easy to get along with. And even though he was a professor and I was a post-doctoral student, with no definite job, he treated me as an equal and so he did with all people in his institute.'

As we have seen, one of Hitler's first measures was a decree excluding 'non-Aryans' from Germany's civil service, and thus the universities. As in many other fields, this was to have a crippling effect on nuclear physics in Germany. Hans Bethe was caught by the law which stated that anyone with one or more Jewish grandparents was disbarred from official work: '...in early '33 the Nazis came in and Heisenberg lost his assistant, Felix Bloch, who had been with him for many years, and so Heisenberg offered me the job and I had to tell him that I was subject to the same Nazi laws as Bloch since my mother was Jewish.'

Like several of the braver German academics, however, Heisenberg continued to show his support for his Jewish friends: 'How did he respond to that?... Well, he invited me to a conference in Leipzig on magnetism which happened to come at that time and we got along very well at that conference, there were lots of people around.'

But for Bethe, like many others, the writing was on the wall:

> I saw immediately that I had to emigrate. Now emigration was easy in 1933, I just left and I went to England for a year and a half. And England was the most hospitable country to German scholars – displaced German scholars – and only gradually America came in also. It was clear to me that I could not look forward to a permanent job in England: there were far too many displaced scholars and too few jobs and the only physicist who got a really steady job was Rudolf Peierls.

The arrival of the Nazis in power also brought personal attacks on scientists like Heisenberg who continued to practise 'Jewish physics'. When Arnold Sommerfeld was due to retire from the chair in physics at Munich University in 1936, he and the University authorities were anxious that Heisenberg, who was continuing to teach at Leipzig, should take over from him. But Heisenberg's refusal to join the Nazi Party and his continued adherence to, and teaching of, the theories of Einstein and Bohr (who was also partly Jewish) gave his enemies among the Nazi physicists powerful ammunition to use against him. In July 1937 Johannes Stark, a Nobel Prize-winning proponent of 'German Physics', published an attack on Heisenberg in *Das Schwarze Korps*, the weekly magazine of the SS, edited by the radical Nazi journalist Günter D'Alquen, describing him as a 'white Jew' whose career had been based on his Jewish friends and their influence. Under these circumstances the Third Reich seemed an increasingly inhospitable place for Heisenberg to practise his physics, and many of his Jewish and foreign friends began to wonder if he would emigrate.

The key to understanding why Heisenberg remained in Germany despite these attacks on him lay in his self-image as a 'Kulturträger'. Heisenberg and many other highly educated Germans genuinely believed that they could ameliorate and soften the impact of Nazism, and that the National Socialists' fiercely 'progressive' modernism must have some positive features. Heisenberg was able to survive the attacks on him because his mother happened to be an old acquaintance of the mother of Heinrich Himmler, head of the SS and one of the guardians of racial purity in the new Germany. Heisenberg's mother approached Himmler's mother and had soon sorted out his problems. Heisenberg later related their meeting thus:

> She said that the elderly Mrs Himmler said immediately, 'My heavens, if my Heinrich only knew of this, then he would immediately do something about it. There are some unpleasant people around Heinrich, but this is of course quite disgusting. But I will tell my Heinrich about it. He is such a nice boy – always congratulates me on my birthday and sends me flowers and such. So if I say just a single word to him, he will set the matter in order.

Himmler's investigation into Heisenberg proceeded slowly, while Heisenberg toyed with the idea of taking a job in New York. Eventually, after a year, the SS decided that Heisenberg was too important for German science to lose him and it was agreed that no further attacks would appear; in return for this, Heisenberg agreed with Himmler that he would continue to teach the 'new physics' without mentioning the names of those responsible for discovering it. Some of Heisenberg's colleagues and friends might have expected him to take a stand against this condition, but he didn't. His daughter, Barbara Blum, argues:

> His decision to stay in Germany was not made lightly, he discussed it with a great many people, and it was actually Max Planck's advice which clinched it, when he told him that it was a terrible time, but that it would pass and that it was important for the people with the right ideas and values to stay [in Germany], live through this terrible time and then help to rebuild things afterwards.
>
> He took this very much to heart and to all intents and purposes it determined his decision, so he had no doubts about it later on.
>
> He felt desperate about not being able to communicate this reasoning to others and get them to understand it.
>
> It weighed on him immensely.

Gerald Holton explains why: 'He went along with that and I'm sure it pained him: more perhaps than some other German physicists, but there wasn't really very much choice. He couldn't have said to Himmler, 'No, I can't accept that condition'. I mean that would have ... been a very big danger, so it was something that he gave in to. He was, as I say, a pragmatist in that sense, with the hope that eventually the good side will win. How this would happen he never explained.'

Heisenberg's inscrutability about the negative effects of Nazi rule lies at the heart of the debate over why the German atomic-bomb project failed. In the summer of 1939 Heisenberg visited the USA to say goodbye to his friends and colleagues before the war that most believed to be inevitable broke out in Europe. As he criss-crossed North America he was urged time and time again by his friends, Americans and German Jewish refugees alike, to stay there and avoid the war; he refused. Hans Bethe spoke to him at this time: 'In 1939 I saw Heisenberg when we both lectured at Purdue University, and we talked briefly about the uranium problem. There was no point discussing the political situation; we knew very well there was general expectation that there would be a war so I didn't talk much about the political situation, but the people at Columbia [University, New York City] did and tried to persuade him to stay in the United States, which he absolutely refused.'

Heisenberg also visited Berkeley, where he had a similar conversation with Robert Oppenheimer and Philip Morrison:

> Well, you know, he talked something about physics and then the conversation in the afternoon began to turn to more political matters and Robert Oppenheimer knew him well, they had been together at the Bohr institute, I believe, [and] began to question him, 'Did he really want to go back? Wasn't it better to stay in America?' And he said, well, no, he couldn't do that. He made it quite clear he was a German, he belonged in Germany and the war might be destructive so he wanted to be there to help, whatever came of it. We tried to point out that it was a bad regime and so on but it was hard to do that to a man who'd already got his mind made up, so when he went away we felt pretty sure he was going to go back, which he did.

In fact by then Nazi Germany had already begun to look into possible military uses of nuclear fission. In April 1939 a conference was called at the Reich Ministry of Education under the chairmanship of Professor Abraham Esau to discuss how a 'uranium burner' might be constructed, and preparations began for obtaining stocks of uranium from within Germany. A second group, led by Professor Paul Harteck, wrote directly to the German War Office: 'We take the liberty of calling to your attention the newest development in nuclear physics, which, in our opinion, will probably make it possible to produce an explosive many orders of magnitude more powerful than the conventional ones.'

A participant of the conference with Professor Esau had been Otto Hahn's deputy, Josef Mattauch. On his return to the Kaiser Wilhelm Institute he discussed the findings of the meeting with several of his colleagues who shared the concern he felt that Nazi Germany might try to acquire an atomic bomb. In Sweden at Christmas, Lise Meitner and Otto Frisch had calculated that the energy released by one atom fissioning would be enough to make a grain of sand jump visibly (which is impressive when one considers that there are about 2,500,000,000,000,000,000,000 atoms in the average grain of sand). Now Siegfried Flügge, one of Mattauch's friends at the Kaiser Wilhelm Institute, decided to flag Germany's interest in nuclear fission to the scientific world. He published an article in which he calculated that:

> One cubic metre of consolidated uranium oxide powder weighs 4.2 tons and contains 3,000 million-million-million-million molecules, or three times as many uranium atoms. As each atom liberates about 180 million electron-volts (about three ten thousands of an erg), in other words three million-millionths of a kilogram metre, a total energy of 27,000 million-million kilogram metres would be liberated. This means that one cubic metre of uranium oxide would be sufficient to lift a cubic kilometre of water (total weight: 10,000 billion kilograms) twenty seven kilometres into the air!

Interestingly, there is a suggestion that Flügge deliberately published this paper in order to alert the rest of the world to Germany's potential military interest in fission. That was certainly the effect it had: within weeks scientists in Britain, the USA, France and elsewhere were taking steps to alert their governments to the possibilities of Hahn's discovery. In the USA, this led to the famous letter of August 1939 from Albert Einstein to Franklin Roosevelt – a letter that was actually largely ignored – while in Britain, a country somewhat more threatened by Nazi Germany, a number of influential physicists, including GP Thomson at Imperial College in London – Britain's foremost scientific 'centre of excellence' – and James Chadwick at Liverpool University, both took their concerns to senior officials.

Hahn's discovery of nuclear fission came at a critical moment in history. A few months earlier the Munich agreement had narrowly averted a European war and bought an extra year of preparation for Britain and France at the expense of Czechoslovakia; but, despite Neville Chamberlain's assurances of 'peace in our time', there was no real confidence that that was what he had achieved. As Europe began to gear itself up for war in the summer of 1939, it appeared that the international nuclear physics community was to be mobilized with it.

CHAPTER SIX

the uranium club

Germany marched into Poland on 1 September 1939, soon after Heisenberg had returned from his farewell tour of the USA. He was not, at that time, expecting to do war work in connection with nuclear physics: he was a reserve soldier in the Wehrmacht's Alpenjäger (mountain infantry) and fully expected to be called up to fight as an ordinary soldier, but fortunately for him, he was not.

In fact one of his assistants, Erich Bagge, had reported to the war office on 8 September, expecting to be sent off to fight as an infantry soldier, only to be met at the War Office in Berlin by two civilians. They turned out to be working for another physicist, Kurt Diebner, who had been placed in charge of the studies by the Heereswaffenamt, a military technical research agency, into nuclear physics, and who wanted Bagge to draw up an agenda for a conference on how nuclear fission might be useful for the German war effort. Bagge suggested to Diebner that Heisenberg would need to be included in any experimental nuclear physics programme. Even though Heisenberg's work had not hitherto been of direct relevance to the production of energy from nuclear fission, he was one of the two or three most respected physicists in Germany, and certainly the most creative and original. Diebner baulked at this: as an experimenter, Heisenberg and the theoretical physicists were anathema to him. Even so, on 16 September 1939, when Diebner's conference assembled in Berlin, an able group of scientists were present, including Otto Hahn, Hans Geiger, Siegfried Flügge, Walther Bothe, Paul Harteck, Gerhard Hoffmann, Kurt Diebner, Bagge and others. Their conclusion was that fission might have a military use, maybe.

Once again Bagge intervened on behalf of his boss: Heisenberg must be invited to join the group. They must all have been thinking of Henry Moseley. In 1912 the young physicist was working on exact methods for measuring the wavelengths of X-rays, and to test Bohr's theories he calculated the nuclear charge for each of a series of elements of medium atomic weight. His results, published in 1913, showed that the positive charge on each nucleus is a whole-number multiple of the negative electrical charge on an electron. In the opinion of his fellow Briton, the respected physicist Frederick Soddy: 'Moseley ... called the roll of the elements, so that for the first time we could say definitely the number of possible elements between the beginning and the end, and the number that still remained to be found.'

But, as an army officer, Moseley's remarkable brain was splattered all across the sand and shingle of Gallipoli by a Turkish sniper. This could not be allowed to happen to Heisenberg: he was Germany's premier theoretical physicist. This time Diebner relented: Heisenberg was mobilized to join the Heereswaffenamt nuclear physics research group. He reported on 26 September 1939 to the Heereswaffenamt in Berlin.

The group of scientists which Heisenberg now joined had begun to jokingly refer to themselves as the Uranverein, the Uranium Club. Although they were now working for the Heereswaffenamt, and despite the fact that the army had taken control of the Kaiser Wilhelm Institute for Physics in Berlin for military research, the majority of them continued to work within their own institutes, where, it was agreed, they would do fission research while continuing to teach. This was precisely the opposite of what eventually happened with America's Manhattan Project, where, as far as possible, the most creative nuclear physicists were concentrated together so that they use their proximity to bounce ideas off each other, and may well have been a contributory factor in the failures of German nuclear research. The very hierarchical structure of the German academic system was most likely to blame: none of the senior physicists was keen to relinquish his title and status. Although Diebner was theoretically in charge of the Uranverein and was able to impose some direction on the research of the individual teams within it, in practice he had little authority over them.

At the first meeting which Heisenberg attended, the proposal had been made that they would build a uranium reactor using natural uranium and a 'moderator' of hydrogen or carbon. Heisenberg was given the task of working out a theoretical basis for this, and work began.

By now serious study into the possible military applications of nuclear fission was taking place in Germany, France and Britain – the three major combatant powers – as well as in the USA (although on a smaller scale). In effect, they all started at the same level: the nuclear physics community was so small that the scientists of each nation knew one another well and knew one another's work intimately, but the start of the war had halted international scientific cooperation and now the scientists were 'on their own'. Until now progress in nuclear physics had been somewhat like a relay race: as each breakthrough was made, the state of the art was passed on like a 'baton', irrespective of international boundaries, but in wartime scientific knowledge suddenly acquired the status of military secrets and the scientists had to work in isolation, without the advantage of their international peers auditing their work. The Uranium Club was confident that Germany had the best nuclear physicists and would therefore achieve the most: in fact they didn't.

The next crucial stride forward in the development of atomic weapons came about more or less by chance and, although it was made by two German scientists, it happened in Birmingham. It had been recognized since the mid-1930s that uranium consisted of two isotopes, one of atomic weight 238 and one of 235. U235 comprised only about 0.7 per cent of any given sample of uranium but would, it was found, fission when it absorbed any neutron at all. U238 will only fission on absorbing a slow neutron and, because of the proportions of the two isotopes in any given sample of natural uranium, a gigantic amount would be needed before there were sufficient U235 nuclei to ensure a chain reaction could take place (estimates, once this became understood, indicated a figure of some fifty tons – certainly too much to be carried in an aircraft). Bohr had argued that a chain reaction in a mass of uranium of this size would not in fact work as an explosive because it could only grow at a moderate rate, and thus the material would heat up, melt and partially evaporate before a violent explosion could take place.

It so happened that Otto Frisch was working at Birmingham University at this time. Although a refugee from Nazism, he was still classified as an 'enemy alien' and, while not interned, he was not permitted to take part in the important war work which was occupying most British academic physicists. Thus he had sufficient free time to pursue his own intellectual curiosities. Early in the spring of 1940 he fell to pondering:

> ...if one could ... produce enough uranium 235 to make a truly explosive chain reaction possible, not dependent on slow neutrons. How much of the isotope would be needed? I used a formula derived by the French theoretician Francis Perrin and refined by Peierls to get an estimate. Of course I didn't know how strongly fission neutrons would react with uranium 235, but a plausible estimate gave me a figure for the required

> amount of uranium 235. To my amazement it was much smaller than I had expected; it was not a matter of tons, but something like a pound or two.
>
> Of course I discussed the result with Peierls at once. I had worked out the possible efficiency of my separation system with the help of Clusius's formula, and we came to the conclusion that with something like a hundred thousand similar separation tubes one might produce a pound of reasonably pure uranium 235 in a modest time, measured in weeks. At that point we stared at each other and realized an atomic bomb might after all be possible.

They took their findings to their department head, Mark Oliphant, an Australian, who passed them on to Sir Henry Tizard, the Rector of Imperial College, who was running a committee advising the government on scientific problems concerned with warfare. Tizard took the report seriously enough to form yet another committee to analyse and examine Frisch and Peierls' results: the MAUD Committee. Crucially, Frisch had demonstrated that the critical mass of uranium 235 was of an order of magnitude which made acquiring it a practical proposition. In some ways this was terrifying: if a Jewish refugee working practically alone in Birmingham could make this calculation, what about the concentrated intellectual powerhouse of the Uranverein? But this was the crucial, key step which the Germans didn't take: although they also recognized that U235 could be used to make an atomic weapon, they did not at this stage make any effort to calculate how much it would require. After some months spent studying the problems of isotope separation – an extremely difficult task because U238 and U235 are chemically identical – and failing to achieve a workable solution, they began to assume that the industrial effort required to acquire sufficient U235 was beyond their capacity.

Instead some members of the Uranverein decided to follow a different route. One of Heisenberg's closest collaborators, Carl Friedrich von Weizsäcker, had been studying the decay of U238. He worked out that in some circumstances U238 could absorb a neutron to create an unstable radioactive isotope, 'U239' (which was in reality element 93, neptunium, with a radioactive half-life of 2.3 days). But what von Weizsäcker saw was that the decay product of element 93 would be element 94, a highly fissile metal. Element 94, which he called 'eka rhenium', has become better known by the name given it by Glenn T. Seaborg, a young American physicist who succeeded in making some in March 1941: plutonium.

Interestingly, the early stages of German wartime nuclear research were in some respects more successful than those in Britain and the USA, but a series of key errors hampered the Germans' efforts throughout the war. Heisenberg's theoretical enquiries into how a reactor should be built had shown the need for a moderator, a substance which would act as a brake on neutrons being emitted from the uranium and slow them to the optimum speed for capture by uranium nuclei. The best substances for this appeared to be hydrogen and carbon. By the summer of 1940 work had begun on a special facility in the grounds of the Kaiser Wilhelm Institute in Berlin – called the Virus House to discourage the curious – where such theories could be put into practice. Throughout 1940 and into 1941 a series of experiments took place at the Virus House and in several other institutions designed to select the most effective moderator. Heisenberg's original papers had pointed towards heavy water as the best moderator, because it had a very low neutron absorption rate, but the immense difficulty of acquiring any had led the

Uranverein to look at other materials as well. The first tests were run with carbon dioxide in the form of dry ice, but these proved inconclusive and instead Professor Walther Bothe turned to graphite, a pure form of carbon.

Graphite should have worked, but the sample materials he used were contaminated with boron, which has a high neutron absorption rate, and it didn't. Instead they would have to use heavy water. The problem with this was that heavy water – water in which the hydrogen element is made up of the heavier deuterium isotope of hydrogen – was extremely difficult to come by. The only industrial-scale source was at the Norsk-Hydro plant at Vemork in Norway, where production amounted to a few hundred litres per year. Heisenberg's studies had shown that several tons would probably be needed. Although the plant was under German control following the occupation of Norway in May 1940 and production could be stepped up, nevertheless there was likely to be a very considerable delay before sufficient heavy water would be available to make a nuclear pile capable of achieving criticality and a self-sustaining reaction.

Thus the Uranverein continued with their experiments at a somewhat leisurely pace, each member probing forward on his allotted task as they attempted to unleash the power of the atom in the service of the Third Reich. In Leipzig, Heisenberg and his assistant Robert Döpel built a small sub-critical reactor with which they were able to measure, for the first time, neutron multiplication after they had introduced a neutron source, but it was a long way from being a self-sustaining chain reaction. (This was probably a good thing: Heisenberg wrongly believed that slow neutron chain reactions – reactors rather than bombs – would be self-stabilizing because the heat of the pile would begin to inhibit neutron capture, and consequently none of the reactors the Germans built contained radiation shielding or any quick means of shutting it down. If they had achieved a chain reaction anyone close to the reactor would have received a fatal dose of radiation poisoning.) There were other problems with the design as well:

> During the experiments there were two accidents and the mechanic who belonged to the theoretical physics institute, Mr Paschen, injured his hand. There was some ... hydrogen in the uranium pile or 'Uranmaschine', and this hydrogen [caught fire], and the whole uranium sphere became very hot... Heisenberg was in the seminar and they called... for Heisenberg. Heisenberg came from the seminar and they called for the fire brigade but the fire brigade had no idea what to do with this very hot pile, so they tried to stop the fire by some blankets and finally they succeeded ... but the contents of this uranium pile ... was destroyed. That means the uranium metal was oxidized and the heavy water was more or less destroyed, so after these accidents the experiments were stopped in Leipzig and further experiments were done in Berlin.

This was not a proto-Chernobyl: the reactor was nowhere near critical. Even so, it reveals a carelessness and amateurism completely at variance with the rigour which was later associated with the Manhattan Project. Meanwhile, as Heisenberg continued to tinker with his reactor design, elsewhere various methods of isotope separation for U235 were tried and failed.

But this picture of limited progress was suddenly overturned. In the summer of 1940 Frits Houtermans, a leftist physicist, was returned to Germany from the Soviet Union, where he had been languishing in jail as the result of the Stalinist terror. Houtermans had been teaching and researching in the

Ukrainian city of Kharkov but had fallen foul of the system and spent three years in prison for a crime that had not actually been committed. On his return to Germany he remained in jail, until the efforts of the chemist Max von Laue, an old friend, secured his release. Banned from taking government work in a university or public institute because of his known political sympathies, von Laue found Houtermans a job working for Manfred von Ardenne, an independent inventor, who owned a laboratory in Berlin and was being funded by the Reich post office to investigate nuclear fission.

Using as his starting point a paper of Bohr's from 1939, Houtermans followed von Weizsäcker's work on element 94 but took it a stage further: he saw that it was likely that chemically separable quantities of element 94 should be produced by a slow neutron chain reaction; in effect he had eliminated the need for enormously difficult isotope separation. He also made the first serious German calculation of the size of the critical mass of U235 – estimating the figure at a few kilograms – and described the fast neutron fission process which would lead to explosive chain reactions in U235.

Houtermans's paper was circulated to the Uranverein in August 1941 and Heisenberg immediately saw that the road was open towards the production of an atomic weapon. The effect of this was to bring the moral qualms that he certainly felt about working for the Nazi government – as the result of their anti-Semitism and aggression – to the forefront of his mind. It is undoubtedly the case that Heisenberg, von Weizsäcker, Houtermans and several of their colleagues were sufficiently opposed to the Nazi regime to be strongly ambivalent about their role in the possible production of an atomic bomb. In doing basic nuclear research Heisenberg appears to have convinced himself that he would be shielding young scientists from having to perform dangerous military service and would be keeping German science alive for reconstruction in the aftermath of the war, which he did not, at this early stage, believe Germany could win.

But by the time Houtermans's results were known to him, it appeared that Germany could not lose. With the major exception of Britain, Hitler had conquered almost all of western Europe, Poland, Yugoslavia, Czechoslovakia and Greece, and from June 1941 had unleashed his Wehrmacht against the Soviet Union with apparently devastating results. Now that Heisenberg knew that an atomic bomb was possible, albeit some way down the line, should he be responsible for placing such a weapon in the hands of the Nazis? It was a question he apparently discussed with both von Weizsäcker and Houtermans; although neither could resolve his dilemma, von Weizsäcker, it is claimed, did suggest that Heisenberg should perhaps discuss the problem with his old mentor and friend Niels Bohr.

There were three objections to this and the first was practical: Bohr was in Copenhagen, and while Germany had occupied Denmark since May 1940, in the repressive atmosphere of the Third Reich, Heisenberg would need a good reason to travel there. The second objection was moral: despite his qualms about the entire project, the Uranverein and its aims were still a major state secret with which Heisenberg had voluntarily associated himself; if he revealed the secret to Bohr, there was every reason to suppose that it might get back to Germany's enemies. The final reason was in some ways personal: Bohr and Heisenberg had been close friends and collaborators, but that had been before Hitler's invasion of Denmark; since then they had had no contact. Heisenberg knew Bohr to be a patriot with a deep love of his country; it is likely that he also knew that Bohr had rejected all the feelers put out to him by the occupying power. Bohr's attitude to Germans in general was one of deep anger and resentment and it was difficult to tell whether he would extend this to his old friend.

The first problem proved the easiest to overcome. Von Weizsäcker's father was the senior civil servant at the Auswärtiges Amt, the German Foreign Ministry, a secret opponent of the Nazis and in a position to engineer an invitation to Heisenberg from the German Embassy in Denmark to give a lecture on behalf of the German Scientific Institute. With this organized, the younger von Weizsäcker wrote to Bohr telling him that he and Heisenberg would be in Copenhagen for the week of 15–19 September 1941 in connection with the lectures (von Weizsäcker was giving one as well) and that Bohr was welcome to attend if he cared to do so. To nobody's particular surprise, no Danes accepted the invitation to hear Heisenberg at Copenhagen's German Scientific Institute, but he was invited instead to repeat his lecture at Bohr's Institute for Theoretical Physics and duly did so.

During his week in Copenhagen Heisenberg visited Bohr's institute on several occasions, but, for all his moral qualms about the war, he nevertheless managed to cause deep offence to many of his Danish contacts. The reason for this was that Heisenberg, although no Nazi, was undoubtedly a German patriot. In September 1941 the Third Reich reached its military high-water mark: the German offensive in Russia was beginning to threaten Moscow and appeared unstoppable, the war seemed as good as won and Heisenberg gave the appearance of being pleased with this. In a guarded way he accepted that the German occupation of western Europe was 'sad', but he argued in favour of the occupation of Poland and Russia, as if he believed the Nazi propaganda about the Germans having some kind of civilizing mission there. This was not what the defeated Danes wished to hear.

But in addition to the formal contacts between Heisenberg and Bohr there was one attempt to renew the friendship between the two. After agonizing for several days Bohr invited his old friend to his home for dinner; if indeed it was Heisenberg's intention to seek advice from Bohr and perhaps pass on a warning to the British (at this stage the USA was still three months away from entering the war and might not have done so at all), then it went disastrously wrong.

The only detailed accounts of the meeting between Heisenberg and Bohr to have emerged so far have come from the Heisenberg side. After dinner, it is said, Heisenberg asked Bohr if he would like to go for a walk, as had been their custom in happier times. As they walked Bohr took Heisenberg to task for his remarks about the desirability of German victory in the east and for Germany's policy of destruction in occupied Poland. Heisenberg's lame response to this was that at least they had not done the same thing in France, which angered Bohr, and he blundered on to say that he still thought defeating Russia would be a good thing.

Heisenberg then suggested that Bohr would be able to achieve greater personal security in occupied Denmark if he could establish some kind of contact with the German Embassy and occupation authorities. In referring to this he probably had in mind the generally anti-Nazi contacts of von Weizsäcker's father, who had engineered his own visit to Copenhagen, but it is clear that Bohr took him to mean that he might improve his own position by actions which were tantamount to collaboration with the occupying power.

Having thus unwittingly succeeded in provoking a mood of extreme hostility and suspicion in Bohr, Heisenberg's supporters say, he now attempted to broach the most important subject on his mind. In effect Heisenberg now asked Bohr if he thought it was right for scientists to work on nuclear fission in wartime. Bohr was shocked: in his isolation in Denmark he had not heard of the new developments that made fast neutron chain reactions possible; he asked Heisenberg if it was possible to use uranium fission to make bombs. Heisenberg apparently told him that yes, it was, that the Germans were working on it, but that it required an enormous technical effort. By now Bohr was hardly listening, or so Heisenberg believed, because Heisenberg claims he went on to say that the construction of bombs would be so difficult that the scientists working on them would have the chance to tell their governments that it would not be technically feasible to bring such a project to fruition in time for atomic weapons to have a role to play in the present war, and that perhaps they, Bohr and Heisenberg, should use their influence on their fellow scientists to stop this situation from coming about. What this really meant was that Bohr should try to stop the British (and presumably Americans) while Heisenberg would do the same for the German side.

Even to Heisenberg, this subsequently came to appear ridiculous and unreasonable. Germany was apparently on course to win the war through conventional armaments alone, and thus to impose a Nazi tyranny on Europe, yet he was asking Germany's enemies, including many German Jews who had already personally suffered under Hitler's rule, to give up work on one weapon that might prove decisive against the Third Reich. Even accepting the kindest interpretation – that of Heisenberg and von Weizsäcker – of Heisenberg's actions, it appears ridiculously naïve.

Heisenberg's version of events is, however, hotly disputed. In the late 1950s a book appeared in which Heisenberg and von Weizsäcker gave their side of the argument. Although Bohr did not publicly respond, he did write a letter which he kept between the pages of his copy, giving his version of events. The Bohr family have decided to embargo the letter until the fiftieth anniversary of Bohr's death, in 2012, but before this decision was made it was seen by several scholars, including Professor Gerald Holton:

> Niels Bohr has not been heard from yet and he has written a letter which I have seen – it was shown to me by the family – and which strongly disagrees with the interpretation that Heisenberg gave to that meeting later... It is not the same to such a degree that Niels Bohr ended up having uncharacteristically written a very harsh letter, by his standards, to object to the story that was given out about moral compunction on the part of the Germans ... to such a degree that he didn't mail it, he put it folded into a book which had Heisenberg's own version – in Robert Jungk's book – to which he took great exception.

In fact, immediately after Heisenberg had left to return to his hotel, Bohr told his family in amazement that Heisenberg had been attempting to pump him for information about nuclear fission bombs. In the wake of the Copenhagen meeting Heisenberg returned to Germany to continue his work. The scientific situation had not been changed by his encounter with Bohr. To put it broadly, by the autumn of 1941 the Germans had successfully dealt with the major theoretical problems of bomb-making but were encountering great technical difficulties in putting them into action. They knew that their two possible routes were either to separate U235 from U238, or to build a reactor which would 'breed' the new element 94; but they had failed to make significant progress in isotope separation and would be reliant on a considerable supply of extremely rare heavy water, even if they managed to perfect a workable reactor design to make plutonium.

It was at this stage that the German grip on eventual victory began to slip. By the beginning of December 1941 leading elements of Field Marshal Fedor von Bock's Army Group Centre had been able to see the early-morning sun glinting on the towers and cupolas of the Kremlin, but a surprise counter-attack organized by the Soviet General Zhukov had thrown them back, and with the Wehrmacht severely overstretched, the entire Eastern Front had threatened to collapse. At the same time Japan's attack on Pearl Harbor had started the war in the Pacific and Hitler had rashly declared war against the USA in support of his ally. By the end of January 1942 the Wehrmacht had suffered 918,000 casualties in Russia, killed, wounded, missing and captured, and suddenly victory no longer seemed secure or even, perhaps, likely.

It was in this atmosphere that the Heereswaffenamt began a review of the prospects for achieving a workable fission weapon within the timeframe of the war. At this point there were a number of groups working on fission weapons within the Uranverein, but it was Heisenberg and his collaborators who made up the theoretical backbone of the project and who thus carried the most influence at a preliminary meeting in December

1941 and a three-day conference in February 1942. What Heisenberg reported then was that a bomb was theoretically possible, but that under the present circumstances, even with a greatly increased level of resources, a practical weapon was still some years away. This has recently been presented as another example of Heisenberg deliberately playing up the difficulties of the bomb project in order to persuade the authorities not to continue with it, but this is a very difficult case to make, and not even Heisenberg himself attempted to do so. In reality it was no more than the truth, and anyone who had worked on the project – not only those who were closely associated with Heisenberg – could only agree.

In February 1942 Fritz Todt, Hitler's personal architect, engineer and Minister for Armaments Production, had persuaded the Führer that victory would be possible only if the German economy switched to total-war mode: efforts must be concentrated solely on war material. Having achieved this, Todt was almost immediately killed in a mysterious air crash and in his place was appointed Albert Speer, another architect and favourite of the Führer. Speer was an exceptionally able man with a clear, technocratic vision of how to proceed in support of the German war effort. Germany was now cut off from most sources of raw materials outside Europe as the result of the British blockade, and was also faced with the ultimate prospect of a war on two fronts as a consequence of the USA's entry into the conflict. Most German strategic thinkers, including Hitler, agreed that this would be difficult, if not impossible, to sustain. Rapid victory against the Russians was therefore essential and, in the interests of short-term results, long-term research projects would have to be cancelled or drastically cut back. Any project which was unlikely to yield concrete results in less than eighteen months would have to go; clearly this included the fission research.

But Speer had been told, over lunch in a fashionable Berlin restaurant, that fission might just be a way for Germany to win the war decisively and he was determined to get a personal briefing from the scientists involved. On 4 June 1942 a high-powered group assembled at the Harnack Haus in Berlin, comprising on one side: Albert Speer, with his advisers Karl-Otto Saur and Ferdinand Porsche (designer of the Volkswagen); General Friedrich Fromm, armaments procurement chief of the OKH; Field Marshal Erhard Milch, Göring's deputy at the Luftwaffe; Admiral Rhein of the Navy; and General Leeb of the Heereswaffenamt. On the other side, together with Heisenberg, were Hahn, Strassman, von Weizsäcker, Bothe, Clusius, Bagge, von Ardenne, Sommerfeld, Harteck and a number of other senior German scientists. As the leading theoretician, it fell to Heisenberg to brief the Reichsminister and military officers.

He gave a broad lecture, covering the progress made by the Uranverein since the beginning of the war and concentrating particularly on prospects for building a uranium reactor. According to Speer, he spoke with some bitterness of the difficulty the Uranverein had encountered in obtaining research materials and resisting the conscription of young scientists into the armed forces. After some discussion Speer asked Heisenberg directly 'how nuclear physics could be applied to the manufacture of atomic bombs'.

Again Heisenberg's response to this has in recent years been dressed up with all kinds of pregnant inner meaning, suggesting that somehow he was deliberately seeking to put the Nazi hierarchy off the idea of building atomic bombs. What

Heisenberg said, according to Speer, was that: '...the scientific solution had already been found and that theoretically nothing stood in the way of building such a bomb. But the technical prerequisites for production would take years to develop, two years at the earliest, even provided that the programme was given maximum support.'

It should be remembered that this was in June 1942. If Heisenberg had been given full resources and backing, and had achieved his timetable, he would have been a full year ahead of the Manhattan Project. Asked how soon the Americans might achieve a bomb, Heisenberg responded that if they worked at maximum capacity they might have a working reactor by the end of 1942 and a bomb perhaps two years later: these predictions were uncannily accurate. The first atomic pile to go critical did so in a disused doubles squash court at the University of Chicago on 2 December 1942; the first bombs were ready by the middle of 1945.

The meeting with Speer effectively marked the end of the German atomic bomb programme. Speer was keen to provide more research funds but, short of a vast, full-scale industrial effort similar to the Manhattan Project, he was not going to acquire atomic bombs for Germany in the foreseeable future. This situation may well have been very much to the liking of the anti-Nazi Heisenberg, but it was certainly beyond his control.

Nuclear research continued, but it was now at a much more modest scale even than hitherto. By contrast, even as these decisions were being made, the US bomb project was beginning to be geared up towards the total effort, involving a workforce of 600,000 men and women and an expenditure of more than two billion dollars. These are figures that could not possibly have been matched by the Germans, even assuming – which on the evidence we have no right to do – that they developed a more efficient and less labour-intensive means of producing U235 or plutonium than the Manhattan Project did. The chief limiting factor now, as Heisenberg and his colleagues in the Uranverein probed towards a critical reactor, was the availability of raw materials. Uranium was not a problem as Germany had control of the Joachimsthaler mines in occupied Czechoslovakia (although the refined metallic form was difficult to acquire, not least because production was disrupted by Allied bombing) but heavy water was still in desperately short supply; and this situation was exacerbated by a series of attacks against the Norsk-Hydro plant.

In February 1943 a team of Norwegian saboteurs trained by Britain's Special Operations Executive succeeded in destroying all the electrolysis cells at Norsk-Hydro, setting back production by several months. This was followed in November 1943 by a US Air Force bombing raid which caused enough damage for the Germans to decide to dismantle the plant and take it back to Germany, where they hoped to resume production. Although they did this, they were never able to bring production back to the same levels that had been achieved in Norway. The final straw came in February 1944 when the Germans sought to move the remaining stocks of heavy water from Norsk-Hydro at Vemork to Berlin. As the valuable containers were being ferried across the 'bottomless' Lake Tinnsjö, charges laid by members of the same group which had blown up the plant detonated, sinking the ferry and consigning the last of the Norwegian heavy water to the depths. Heisenberg had estimated that to build a critical reactor would require five tons of heavy water; in fact they didn't have half of that amount.

The fear of what Heisenberg and his colleagues might achieve had been a driving force behind the Manhattan Project and the fact that the Uranverein had largely given up in their attempts to build an atomic bomb did not percolate through to Allied intelligence until the end of the war. With the situation becoming more and more favourable for Allied operations in 1944, a scheme was launched to try to collect as much information as possible on German nuclear research by sending Manhattan Project personnel into the field in an attempt to round up the key scientists in occupied Europe as they were overrun. At the same time consideration was given to plans to kidnap or assassinate Heisenberg, identified by many of his former friends as the most likely leader of any German bomb project.

The mission to uncover the secrets of German nuclear research was code-named ALSOS and commanded by Lieutenant Colonel Boris Pash of US Army G-2 Security. Having established a base in London, he led teams of soldiers and scientists that moved with the foremost elements of the Allied armies, seizing documents and interviewing scientists of the occupied countries as they went, building an increasing dossier of evidence that the Uranverein had got nowhere with its bomb project. Documents captured in Strasbourg at the beginning of 1945 indicated that the senior German physicists were mostly working in southern Germany, away from the heaviest bombing. On 17 April 1945 an ALSOS team led by Colonel John Lansdale discovered the German stockpile of uranium in an open-sided shed at a factory in Stassfurt near Magdeburg; on 23 April, Pash, Lansdale and a few of their colleagues forced the lock on a doorway leading into a cave in the village of Haigerloch. There they found a cylindrical pit lined with graphite and covered with a heavy metal lid: the remains of Heisenberg's last experimental Uranmaschine. Michael Perrin, a British scientist with the group, quickly realized that it could never have gone critical, there was no shielding and the radiation would have killed everyone in the cave.

So why did Heisenberg and the Uranverein fail to build a bomb? In the years following the Second World War the story grew up that it was the result of basic errors and incompetence by the German physicists, whose ability to perform good science was crippled by the interference of Nazi doctrine in their work. A more realistic assessment of the achievements of the Uranverein shows that this was not the case, and historians sympathetic to Heisenberg and his colleagues have been suggesting more recently that the reason that so little progress was made lay in the moral qualms that Heisenberg and his closest colleagues felt about placing an atomic weapon in the hands of the Nazis. The reality is that neither is an accurate representation of the truth. There is no doubt that Heisenberg, von Weizsäcker and other senior members of the Uranverein did have strong doubts about the wisdom of developing nuclear weapons for the Nazis, but, as Heisenberg himself consistently maintained in the years following the war and up until his death in 1976, they were ultimately spared the choice about whether to proceed by the lack of available resources.

Even so, there is no question that German nuclear research fell behind the Allies at both the theoretical and the technical level. Immediately after the war many members of the Uranverein were interned at Farm Hall for a period of six months, during which, as we have seen, the Americans dropped their uranium bomb on Hiroshima and their plutonium bomb on Nagasaki. Hans Bethe recalls: '...at Farm Hall, where he and other physicists were interned at the end of the war, Heisenberg was asked to tell the other physicists how the Allies could possibly have made a bomb and so he gave a talk which is preserved in the Farm Hall records and this talk was completely wrong. Heisenberg explained very carefully how a reactor works, he thought the bomb would work the same way, so he really had no idea how a bomb would work.'

From this, Bethe concluded that it was probably correct to assume that Heisenberg hadn't really wanted to build a bomb or that, at least, he assumed that the resources necessary were so far beyond the German capacity during the war that it was not worth trying to find out how to do so. Bethe had no illusions about how much effort it had been for America, as well as the effects of the Uranverein's technical failures:

> Well, the Germans decided at the very beginning that the isotope separation of uranium was beyond their capacity and if you have a look at the big establishment at Oak Ridge, Tennessee, you can see the Germans probably couldn't have done it even if they had wanted to. In addition, by the time it became right Germany was being bombed and any installation like Oak Ridge would surely have been bombed to pieces before it was finished, so the Germans decided at the beginning they would not try to get the active isotope, 235 of uranium.
>
> But they knew that neutrons absorbed by uranium would give plutonium: they didn't call it plutonium yet but that doesn't matter. And they figured out from Bohr's theory that plutonium would be just as good as uranium 235 in making a bomb but, of course, they never got there since their reactor never got started and only after they had a reactor could they even think of making plutonium. They thought, having the idea

> of plutonium, that the way to a bomb was easy and they would have been surprised how difficult it is even once you have a reactor.

It was also true that the German scientists didn't share the terrible fear faced by the Manhattan Project that their enemy might achieve a bomb first. Heisenberg's assumption that the Uranverein couldn't build a practical bomb in wartime extended to the Americans. Bethe again:

> Well, Heisenberg thought that he was ahead in getting a nuclear reactor started and that he could impress the Allies with the amount of progress they had made on the nuclear reactor, he didn't know that in 1942 already Fermi had achieved a nuclear chain reaction and in 1945 Heisenberg was at best halfway to a nuclear chain reaction.
>
> The Germans thought they were the ... best scientists in the world and that no other country could possibly do anything better than they did. Well, I think this was a common belief in Germany: after all Heisenberg had discovered quantum mechanics; Schrödinger who was an Austrian had put it in a form that one could use it easily. It was a German invention and therefore they thought they were the leaders of the world.

conclusion

For most historians of the Third Reich one of the most surprising and shocking aspects of the whole era is how little protest there was at the extreme measures that Hitler's government began to take against their opponents – real and imagined – almost from the start. In many ways Hitler has become a caricature of evil: a hysterical, demonic and murderous figure from our worst nightmares, but in reality he was a populist politician. He was certainly lacking in the moral restraints which we expect to see in normal members of society, but as a politician he was a calculating and cunning operator, careful to stay within the boundaries of what the public were prepared to accept; at least until he had prepared and propagandized them to go further. This is well illustrated by the path that he followed which led to compulsory sterilization, euthanasia and finally genocide.

As we have seen, sterilization was first adopted in some parts of the USA around the start of the twentieth century, and other countries were not far behind. Germany did not begin sterilization until 1933, and this was in a climate in which a widely accepted 'science' claimed to be able to determine an order of rank among individual people and between races. The prejudices on which eugenics was based were almost universal among the developed nations: it would have been no more unnatural or odd for a convinced socialist, or a Christian, to regard an African as a member of an inferior race than it would have been for an arch-conservative. Nor was it unusual for a rich and well-educated member of the upper class to regard himself as morally superior to a member of the working class.

The introduction of compulsory sterilization, then, was a comparatively small step for the Nazis to take, one which had been introduced elsewhere without great protest, and discussed at length in Germany even before they came to power. It is also worth pointing out that in Sweden 63,000 people, mostly women, were sterilized between 1934, when the policy was introduced, and 1975, when it was halted. For most of this time Sweden was ruled by liberal, Social Democrat-led, coalition governments. Although sterilization was supposed to be voluntary, this was only true in a small minority of cases. The victims were selected because they were supposedly 'asocial' although in reality many were of gypsy descent.

In fact in Germany the sterilization of the 'genetically inferior' took on a ritualistic aspect that stood outside the boundaries of any therapeutic effect it might have, either on the individuals themselves or on the gene pool of the German race. Almost all of the victims of sterilization who were confined in asylums or hospitals for the incurable were living in conditions of sexual segregation: they had no realistic opportunity to form relationships which might take on a sexual character, unless they were expected to start copulating like animals; an unlikely prospect under the degree of supervision and regulation they endured. This reflects the reality that sterilization was the answer to a problem which did not actually exist.

The pseudo-science of eugenics which gripped the civilized world in the first half of the twentieth century is strikingly similar to the current concerns over the environment, ecology and overpopulation (which may, in itself, be simply a disguised version of eugenics). Although a number of important World Environmental Summits have taken place to address environmental issues, sponsored by the governments of many different countries as well as the United Nations, there is little real scientific consensus over what the environmental problem is, how it can be tackled or whether there is indeed a problem at all. Certainly the environment is changing, but few would dispute that the environment has always changed. Less than a thousand years ago Greenland supported a colony of Viking settlers who lived on crops they were able to grow in a relatively temperate climate. The Greenland settlers starved to death when their climate cooled and their crops failed for several years in succession. Three hundred years ago London held regular winter 'Frost Fairs' when the waters of the Thames froze solid for weeks on end and night-time temperatures dropped below 0° Fahrenheit. Equally, however, there is plenty of evidence of a time when London enjoyed temperatures we would now describe as tropical.

To make this point is not to say that environmentalists are necessarily wrong, but that the scientific arguments that underpin their case are by no means proven. Although there may be a rise in global temperatures – a 'fact' which remains disputed – it is not necessarily the case that this is caused solely or largely by environmental pollution. It is entirely conceivable that any rise in temperature which is taking place is purely cyclical. Pollution is certainly responsible for some other, health-related, problems and is probably a factor in global warming, if that does actually exist; but can we say more than that?

Looking back to pioneering studies in eugenics, we should compare them with what people say now about the environment. In 1913 the anthropologist Eugen Fischer published his book *Die Rehobother Bastards und das Bastardisierungsproblem beim Menschen*, an account of mixed-race children in the then German colony of South West Africa. These children suffered discrimination and rejection by both the black and white peoples in Namibia. They were a social problem, but in reaching his conclusion, which was, in part, that the mixing of European peoples with 'inferior races' resulted in 'spiritual and cultural degeneration', Fischer was drawing inferences that could not be justified by his research.

We would nowadays argue that Fischer had identified a problem that was neither genetic nor hereditary in nature: in fact we know categorically that he was not scientifically equipped to make such a statement. DNA was not discovered and explained until after the Second World War. Instead any problems faced by the 'Rehoboth bastards' were social and environmental. But his work was enormously influential. In the 1930s it was used as the basis for sterilization of a small group within Germany – the so-called 'Rhineland bastards'. These were children of German mothers who had been fathered by African and Arab soldiers serving in the French army of occupation in Germany in the 1920s. There were so few of them, no more than 500, that they had been overlooked when the compulsory sterilization law was drafted in 1933. But, according to Benno Müller-Hill:

> A meeting of 'Workgroup II of the Expert Advisory Council for Population and Race Policy' was convened on 11 March 1935. The topic for discussion was the sterilization of coloured children... Three possible approaches were considered: widening of the scope of the law, 'export', i.e. deportation, and compulsory sterilization without changing the law. In 1937 a decision was handed down from the Chancellery of the Reich, but was never put on record: compulsory sterilization without any basis in law.

The point about Fischer is that his scientific methodology was considered more or less, impeccable and was also subjected to peer review. The problem is that his peers shared his prejudices; and in the longer Nazi context, the state supported them. Most real National Socialists were not scientists and had very little understanding of it, but they believed their political views on race to have a scientific basis. The eugenic order of rank, supported by IQ tests and puzzling measurements of brain size, was seen to be a product of scientific method; therefore they must be right.

In a slightly different context, the American philosopher of science Thomas S Kuhn has argued that science proceeds by revolutionary change: that scientists construct 'paradigms' within which they work – even though their data often contradicts those paradigms – until another scientist comes along with a new theory which overturns the original paradigm and is better able to explain the experimental data.

A good example of this can be drawn from astronomy. The Ptolemaic astronomical system, devised in the two centuries before and after the birth of Christ, held the Earth to be at the centre of the universe but nevertheless provided a set of mathematical formulae which were able to predict with reasonable accuracy the positions of the stars and the planets, in their assumed orbits around the Earth, for many hundreds of years. As time passed, however, the Ptolemaic predictions became increasingly inaccurate. The two responses to this were either that they were wrong, or that they needed adjustment. Working within the Ptolemaic paradigm, adjustments could be made to the orbits of the stars and the planets – these were actually visualized as small circular excursions from their normal orbits, called epicycles – but as more and more epicycles were added, to explain the increasing inaccuracy of the original model, the more and more unlikely it seemed that the original model was correct. This is what Kuhn describes as a scientific crisis. The eventual response to Ptolemy's astronomy was Copernicus's theory that perhaps the Earth was orbiting the Sun. No doubt much forehead slapping ensued.

There was nobody to do this in the Third Reich – or Sweden or the USA – when scientists asserted that allowing people with 'hereditary diseases' to continue to breed would do damage to the race and was at the root of a lot of social problems. None of the proponents of sterilization claimed that it would work instantaneously to cure the problems of society. Hitler himself is on record as believing that it might take 600 years to breed these faults out of the German race. There was a consensus that there was a problem and that the solution was obvious:

> From the present point of view, sterilization does not appear to be a suitable instrument for actual preventative use but, from the state of knowledge at that time, it appeared to provide a realistic chance, through mass, and also forced, sterilization, of eradicating diseases and handicaps in the next generation.
>
> When we look at the indications that are listed in the law on preventing the birth of offspring with hereditary diseases, which came into force in 1934, one sees that for one thing it concerned mental handicaps. The main focus here was on so-called feeble-mindedness, and they got around the problem of heredity by talking about innate feeble-mindedness; not inherited feeble-mindedness: innate feeble-mindedness.
>
> Then it was concerned with the area of mental illness, especially schizophrenia and epilepsy, but it was also concerned with social deviation; you can see that from the fact that severe alcoholism was included in the law, and the diagnosis of innate feeble-mindedness provided the opportunity for including socially deviant people in the sterilization programme, say backward children or those with a criminal record.

> Sterilization was widely accepted, both in the medical profession as well as in society. That extends, when one looks at the political spectrum, from the extreme right to the extreme left. It also includes the churches – the Protestant Church took a very active part in the debates and was in favour of sterilization and in the Home Mission institutions it also took a very active part in the sterilization programme. Also in the Catholic Church there were voices in favour of eugenics, and in support of sterilization, at least voluntary sterilization.
>
> I would just mention Hermann Muckermann, Head of Department at the Kaiser Wilhelm Institute for Anthropology, Human Heredity Biology and Eugenics, who was both a Jesuit priest and a scientist. Such Catholic scientists were brought back into line by papal authority, but I would say that even in Catholic circles there was sympathy for sterilization.
>
> When I have to explain it to [my students] I always say: 'Anyone who was against sterilization at that time, for reasons of principle, was regarded in about the same way as opponents of vaccination today who cite religious reasons.'

Within this paradigm, no 'crisis' had arisen and none was expected to. We can certainly argue, without stretching the analogy beyond breaking point, that we are in the same position on global warming. It may well be the case that if the global climate warms up by a few degrees it will cause catastrophic damage in low-lying island states, but we cannot yet be sure that we are experiencing global warming, or that it is caused by problems which can be addressed by human intervention: nevertheless, serious steps are being taken to counteract it.

The origins of the euthanasia programme were different. There was no suggestion that the incurably ill were the root of a serious social problem, but they could certainly be claimed to be part of an economic problem:

> The position of psychiatry in Germany in the interwar period is defined by two quite essential viewpoints.
>
> One is that for the first time it appeared possible to develop truly effective forms of treatment. In the twenties, so-called more 'active' treatment was developed, that was a very comprehensive programme of occupational therapy, which had an absolutely revolutionary effect in the institutions. In addition, there was a new concept of open care – that is, the institutions and semi-outpatient systems were opened up and mentally ill people were discharged into society. In the thirties there came the so-called shock therapies, insulin and cardiazole shock, at the beginning of the forties, electric shock, and all these forms of treatment appeared to open up entirely new possibilities: for the first time in psychiatry, it appeared possible that people could actually be cured, or their conditions improved to the extent that they were capable of working and enjoying themselves.
>
> That was the one side. The other side was that the institutions had been in a severe crisis since the First World War, an overcrowding crisis. The number of people in psychiatric institutions was constantly increasing, while the number of beds stagnated: that is, the institutions became ever fuller. And the proportion of people who were in institutions for long periods, the chronically ill, was constantly increasing; while the proportion of acutely ill decreased.
>
> And that meant that the conditions in mental hospitals and nursing homes became virtually catastrophic, especially after the outbreak of the world economic crisis [of the early 1930s]. From these two points of view there arose an absolutely fatal question, which the medical experts, who then finally cooperated with the euthanasia, asked themselves: If the

resources, the space, the money, the personnel, the equipment are not sufficient to serve all patients properly, shouldn't we get rid of the encumbrances, that is the chronically ill, the mentally handicapped, who we can't reach with our therapeutic treatments anyway? Shouldn't we simply get rid of them at a stroke, in order to make the way free for progressive psychiatry, and thus be able to help as many people as possible?

And from this consideration of loving kindness these doctors came to the point that they made themselves available to kill their own patients. There was resistance, there were individual doctors who were not prepared to take part in the action. They were left alone, there was no pressure at this level, I should like to emphasize that. There was a lot of partial resistance on the part of doctors in the mental hospitals and nursing homes, that is, there were attempts to prolong the action. They sometimes also refused to complete the report forms that were the basis for selection, and so on. But one has to say that these forms of partial resistance also always included partial collaboration. The idea that one could divide the patients in the institutions into those who were curable and those who were incurable, those who were productive and those who were unproductive, that idea was very widespread, not only among the doctors who cooperated in the euthanasia, but also among the doctors who were against it.

I'll give an example: in the von Bodelschwing institution in Bethel, towards the end of 1940, when they could foresee that the attitude of refusal that had been adopted until then could no longer be maintained, they asked themselves whether the best way to proceed wouldn't be to select the patients themselves and then present the completed selection to the euthanasia apparatus. And that is exactly what they did.

So they divided their own patients into different categories, and presented this categorization to the Medical Commission that came to Bethel from Berlin. And it was accepted. They wouldn't have been able to do that if they hadn't had it in their heads that there is life unworthy of life.

This is an idea that is interesting, because although no serious commentator would advocate involuntary euthanasia in the same manner that it was carried out in Germany in 1940 and afterwards, nevertheless there is a strong lobby to allow so-called mercy killings in many advanced countries, and the Netherlands actually passed a law to allow this in November 2000. Current proponents of euthanasia argue that doctors should be allowed, without fear of prosecution, to accede to the wishes of their patients to be killed with the minimum amount of pain when the remainder of their life will be unendurable. It is possible to say that it has been routine in modern medicine for doctors to shorten the last few hours of a patient's life by administering an overdose of a therapeutic drug. Less usually, when, for example, a patient has developed a very high tolerance for, say, morphine during the course of their treatment, doctors have occasionally been observed to give a non-therapeutic drug in order to hasten an already terminal situation.

However, there is a problem of degree – where do we lay the boundary where a person's life is unendurable? There have been high profile cases where the relatives and doctors of someone who has been in a permanent vegetative state for several years and seemingly will never make a recovery, have obtained the permission of the courts to terminate that person's life. Although the patient may be able to breathe unassisted, they are dependent on round-the-clock nursing, and, as far as anyone can ascertain, has no consciousness of the world around them. Clearly, the circumstances of each patient are subject to close scrutiny and the decision to end the suffering of such people is an extremely complex and difficult one. However, the fact remains that such decisions are a step beyond a doctor making someone's last hours that bit more comfortable – there is a point where it may be considered that when the chances of recovery are negligible, it seems more humane to end someone's life than to continue it. Leaving aside the undeniable horrors of the circumstances in which patients were killed by the Nazis: the lack of consultation of the patients' families, the methods used to end their lives, and, not least, the fact that many more people were killed whose condition could not be described as incurable, the practice of certain forms of euthanasia in modern society makes it more difficult for us to utterly dismiss the actions of all the Nazi doctors and nurses as denoting a complete reversal of their role as carers and of being motivated by a kind of evil that is incomprehensible. Some of the victims of German euthanasia may have been in a similar condition as people in a permanent vegetative state whose families have obtained the right to end their life. It is uncomfortable for us to contemplate the possiblity that some of those doctors and nurses who were against the process and had to be coerced into making their selection, based their decisions on factors that are still pertinent and surrounded by controversy today.

Part of the problem that underlay the involvement of German doctors in sterilization and euthanasia came from the Hippocratic Oath, which at that time was sworn by doctors as they qualified and which acted as a basis for medical ethics. The difficulty is that the oath is somewhat out of date.

I SWEAR by Apollo the physician, and Aesculapius, and Health, and All-heal, and all the gods and goddesses, that, according to my ability and judgment, I will keep this Oath and this stipulation to reckon him who taught me this Art equally dear to me as my parents, to share my substance with him, and relieve his necessities if required; to look upon his offspring in the same footing as my own brothers, and to teach them this art, if they shall wish to learn it, without fee or stipulation; and that by precept, lecture, and every other mode of instruction, I will impart a knowledge of the Art to my own sons, and those of my teachers, and to disciples bound by a stipulation and oath according to the law of medicine, but to none others. I will follow that system of regimen which, according to my ability and judgment, I consider for the benefit of my patients, and abstain from whatever is deleterious and mischievous. I will give no deadly medicine to any one if asked, nor suggest any such counsel; and in like manner I will not give to a woman a pessary to produce abortion. With purity and with holiness I will pass my life and practice my Art. I will not cut persons labouring under the stone, but will leave this to be done by men who are practitioners of this work. Into whatever houses I enter, I will go into them for the benefit of the sick, and will abstain from every voluntary act of mischief and corruption; and, further from the seduction of females or males, of freemen and slaves. Whatever, in connection with my professional practice or not, in connection with it, I see or hear, in the life of men, which ought not to be spoken of abroad, I will not divulge, as reckoning that all such should be kept secret. While I continue to keep this Oath inviolate, may it be granted to me to enjoy life and the practice of the art, respected by all men, in all times! But should I trespass and violate this Oath, may the reverse be my lot!

National Socialist doctors could and did argue that this was a simple curio which could be discarded in the modern age. As Michael von Cranach explains:

> ...all had taken the Hippocratic Oath, but I think they interpreted it differently. As we said before, the focus of attention changed. It was not, any more, the individual patient who was the focus of attention of the doctors, but it was the health of the nation, or the health of the race, as they said... And so they always thought they were acting as good doctors,

> because they were perhaps producing now brief suffering of patients, but, in the long term, they would create a healthy ... race.

This leads on to human experimentation. In December 1946 the USA began what was known as the 'Doctors' Trial' at the US Military Tribunal at Nuremberg. Twenty-four doctors accused of involvement in human experimentation on concentration camp inmates and in the sterilization and euthanasia programmes, were tried for their crimes. In the event, some eight were acquitted in large measure because they were able to shift a great deal of the blame – particularly in the high-altitude and low-temperature tests – on to Rascher, who by then had been executed, as an inmate at Dachau, for his role in his wife's child abductions. They argued, sufficiently well to convince the tribunal, that they had attempted to restrain Rascher. Among those convicted, however, was Karl Brandt, Hitler's surgeon and one of those originally selected to organize the T-4 euthanasia apparatus: he received a death sentence and was hanged on 2 June 1948.

One aspect of the defence adopted by many of those being tried at Nuremberg was that human experimentation was commonplace and widely practised in the USA. This was, of course, true, the crucial difference being that ethical human experimentation is only ever carried out with the consent of the subject. The Nazi doctors had used their experimental 'guinea pigs' as objects: commodities to be used and destroyed at a whim. In giving its judgement, the tribunal at Nuremberg promulgated the 'Nuremberg Code', which outlined the basic ethical guidelines to be used in experimenting on humans:

Permissible Medical Experiments

The great weight of the evidence before us is to the effect that certain types of medical experiments on human beings, when kept within reasonably well-defined bounds, conform to the ethics of the medical profession generally. The protagonists of the practice of human experimentation justify their views on the basis that such experiments yield results for the good of society that are unprocurable by other methods or means of study. All agree, however, that certain basic principles must be observed in order to satisfy moral, ethical and legal concepts:

1. The voluntary consent of the human subject is absolutely essential. This means that the person involved should have legal capacity to give consent; should be so situated as to be able to exercise free power of choice, without the intervention of any element of force, fraud, deceit, duress, over-reaching, or other ulterior form of constraint or coercion; and should have sufficient knowledge and comprehension of the elements of the subject matter involved as to enable him to make an understanding and enlightened decision. This latter element requires that before the acceptance of an affirmative decision by the experimental subject there should be made known to him the nature, duration, and purpose of the experiment; the method and means by which it is to be conducted; all inconveniences and hazards reasonably to be expected; and the effects upon his health or person which may possibly come from his participation in the experiment. The duty and responsibility for ascertaining the quality of the consent rests upon each individual who initiates, directs or engages in the experiment. It is a personal duty and responsibility which may not be delegated to another with impunity.

2. The experiment should be such as to yield fruitful results for the good

of society, unprocurable by other methods or means of study, and not random and unnecessary in nature.

3. The experiment should be so designed and based on the results of animal experimentation and a knowledge of the natural history of the disease or other problem under study that the anticipated results will justify the performance of the experiment.

4. The experiment should be so conducted as to avoid all unnecessary physical and mental suffering and injury.

5. No experiment should be conducted where there is an a priori reason to believe that death or disabling injury will occur; except, perhaps, in those experiments where the experimental physicians also serve as subjects.

6. The degree of risk to be taken should never exceed that determined by the humanitarian importance of the problem to be solved by the experiment.

7. Proper preparations should be made and adequate facilities provided to protect the experimental subject against even remote possibilities of injury, disability, or death.

8. The experiment should be conducted only by scientifically qualified persons. The highest degree of skill and care should be required through all stages of the experiment of those who conduct or engage in the experiment.

9. During the course of the experiment the human subject should be at liberty to bring the experiment to an end if he has reached the physical or mental state where continuation of the experiment seems to him to be impossible.

10. During the course of the experiment the scientist in charge must be prepared to terminate the experiment at any stage, if he has probable cause to believe, in the exercise of the good faith, superior skill, and careful judgment required of him that a continuation of the experiment is likely to result in injury, disability, or death to the experimental subject.

Of the ten principles which have been enumerated our judicial concern, of course, is with those requirements which are purely legal in nature – or which at least are so clearly related to matters legal that they assist us in determining criminal culpability and punishment. To go beyond that point would lead us into a field that would be beyond our sphere of competence. However, the point need not be laboured. We find from the evidence that in the medical experiments which have been proved, these ten principles were much more frequently honoured in their breach than in their observance. Many of the concentration camp inmates who were the victims of these atrocities were citizens of countries other than the German Reich. They were non-German nationals, including Jews and 'asocial persons', both prisoners of war and civilians, who had been imprisoned and forced to submit to these tortures and barbarities without so much as a semblance of trial. In every single instance appearing in the record, subjects were used who did not consent to the experiments; indeed, as to some of the experiments, it is not even contended by the defendants that the subjects occupied the status of volunteers. In no case was the experimental subject at liberty of his own free choice to withdraw from any experiment. In many cases experiments were performed by unqualified persons; were conducted at random for no adequate scientific reason, and under revolting physical conditions. All of the experiments were conducted with unnecessary suffering and injury and but very little, if any, precautions were taken to protect or safeguard the human subjects from the possibilities of injury, disability, or death. In every one of the

> experiments the subjects experienced extreme pain or torture, and in most of them they suffered permanent injury, mutilation, or death, either as a direct result of the experiments or because of lack of adequate follow-up care.
>
> Obviously all of these experiments involving brutalities, tortures, disabling injury, and death were performed in complete disregard of international conventions, the laws and customs of war, the general principles of criminal law as derived from the criminal laws of all civilized nations, and Control Council Law No. 10. Manifestly human experiments under such conditions are contrary to 'the principles of the law of nations as they result from the usages established among civilized peoples, from the laws of humanity, and from the dictates of public conscience.'

Ironically, while the trial was taking place other German doctors who had been connected with some of the worst experiments were being employed by the USA. The most senior of these was Dr Hubertus Strughold, the Luftwaffe's most senior specialist in aviation medicine. Strughold had participated in and contributed to the two medical conferences which had discussed Rascher's findings in the low-pressure and freezing experiments and had almost certainly been fully aware of what their consequences had been, but the rubble of post-war Germany was no place to do research on high-performance flight. The most talented of the German aviation doctors, under the leadership of Strughold, were collected by the Americans and spirited off to institutes on military bases all over the USA. Strughold went to San Antonio, Texas, to run the School of Aviation Medicine at Randolph Air Force Base.

The surviving scientific results from the Dachau experiments ended up in a document prepared for the Nuremberg Trials. The cold-water results survived intact, as we have seen, and are quoted in research papers even today. Much of the high-altitude data seems to have been destroyed, which is perhaps why Strughold had the experiments repeated after the war at Randolph on volunteer subjects. Strughold held exactly the same position in cold-war America as he had in Nazi Germany. He was head of his own institute: the pre-eminent Dean of Aviation Medicine. He was responsible for supervising the medical aspects of the USA's rise to its position of overwhelming air supremacy, as well the conquest of space (he designed John Glenn's spacesuit) – and in particular the moon – and a genial figure everyone looked up to. Texas had a Hubertus Strughold Day. There was a prestigious Strughold Award. There was a Strughold Library and a Strughold Institute. Only gradually did an opposing view prevail, and the honours were withdrawn. Nevertheless, the opinion in the US space industry is that, but for Strughold, the Americans would have been ten years behind in the space race and the first man on the moon would have been a Russian.

As we have seen, the failure of the German nuclear programme can be traced to technical mistakes made early on combined with the failure of the Uranverein to make some of the basic calculations which would have led them down the right path. As we have also seen, the leading scientist of the German bomb project, Werner Heisenberg, while indicating that he had had moral qualms about his involvement in building an atomic bomb for Hitler, has never maintained anything other than that he was spared having to make the choice because Germany would not have been in a position in wartime to follow up any successful findings. This is a strictly true interpretation of the facts. There is no question that Germany could have mustered resources equivalent to the Manhattan Project while fighting a war against both the Allies and the Soviet Union. We have seen that Allied air power eventually overwhelmed Germany's air-defence system, notwithstanding the Germans' ability to field remarkably advanced fighter aircraft, and by mid-1944, which is the earliest one can imagine a major German atomic bomb project getting underway, the Allies were able to bomb at will. It is inconceivable that isotope-separation plants on the scale of Oak Ridge, for example, could have survived unnoticed. The suggestion that Heisenberg and the Uranverein chose not to build a bomb for moral reasons has no basis in fact.

But, fundamentally, the German bomb project was crippled by the same factor which led to sterilization, euthanasia and human experiments: racism. By expelling the Jews from their university posts and forcing them overseas, Germany lost the services of some of the greatest minds in physics – people like Otto Frisch and Rudolf Peierls – who were crucially important in building the atomic bomb for potential use against Germany.

Science in the Third Reich achieved remarkable successes in a number of fields – aerodynamics and rocketry spring to mind – but the political system ensured that there was a moral price to pay in almost all of them. The astonishing V-2 rocket, for example, could only be built with the use of hordes of slave labourers working literally at gunpoint; the extraordinary Me163 could only be flown because Rascher had killed hundreds of concentration camp inmates in low-pressure experiments. And in the aftermath of the war this corruption continued to hang over the practice of science in Germany. Germany showed high rates of smoking because, at least in part, people associated anti-smoking campaigns with the Nazi era and were less inclined to take heed of them. German scientists were reluctant to become involved in the human genome project because of the associations that genetic research carried in Germany.

Nazism, like communism, shows us that it is futile and dangerous to attempt to apply simple solutions to complex problems and that issues of ethics apply just as strongly to scientific research as they do in other walks of life. A German chemist, Kurt Mendelssohn, who was forced to flee Germany by Nazi persecution, recalled the day of his arrival in England: 'When I woke up the sun was shining in my face. I had slept deeply, soundly and long – for the first time in many weeks... I had arrived in London and gone to bed without fear that at 3 a.m. a car with a couple of SA men would draw up and take me away.'

The list of names of those who fled Germany because of Nazi persecution of the Jews includes eleven people who had won or would win the Nobel Prize. Few ever cared to resettle there.

BIBLIOGRAPHY

George J Annas and Michael A Grodin, The Nazi Doctors and the Nuremburg Code (Oxford University Press, Oxford, 1992)

Wolfgang Benz, The Holocaust: A Short History (Profile, London, 2000)

Michael Burleigh, The Third Reich: A New History (Macmillan, London, 2000)

I C B Dear and M R D Foot, The Oxford Companion to the Second World War (Oxford University Press, Oxford, 1995)

Ute Deichmann, Biologists under Hitler (Harvard University Press, Cambridge, Massachusetts, US, 1996)

Henry Friedlander, The Origins of Nazi Genocide (University of North Carolina Press, Chapel Hill & London, 1995)

Robert Harris and Jeremy Paxman, A Higher Form of Killing (Chatto & Windus, London, 1982)

Heinz Hoehne, The Order of the Death's Head (Secker & Warburg, London, 1969)

David Irving, Hitler's War (Hodder & Stoughton, London, 1977)

David Irving, The Virus House (William Kimber, London, 1967)

Brian Johnson, The Secret War (Arrow, London, 1979)

Michael H Kater, Doctors under Hitler (University of North Carolina Press, Chapel Hill, US, 1989)

Ian Kershaw, Hitler 1889–1936: Hubris (Allen Lane, Penguin Press, London, 1998)

Ian Kershaw, Hitler 1936–1945: Nemesis (Allen Lane, Penguin Press, London, 2000)

Ian Kershaw, The Nazi Dictatorship (Arnold, London, 2000)

Deborah Lipstadt, Denying the Holocaust (Penguin, London, 1994)

Benno Müller-Hill, Murderous Science (CSHL Press, New York, 1998)

Thomas Powers, Heisenberg's War (Jonathan Cape, London, 1993)

Robert N Proctor, The Nazi War on Cancer (Princeton University Press, Princeton & Oxford, 2000)

Peter Singer, Rethinking Life and Death (Oxford University Press, Oxford, 1995)

Hugh Trevor-Roper, Hitler's Table Talk 1941–44 (Phoenix, London, 2000)

Adrian Weale, Eyewitness: Hiroshima (Robinson, London, 1995)

INDEX